Optics: Theory and Applications

Optics: Theory and Applications

Carl MacKenzie

STATES
ACADEMIC PRESS
www.statesacademicpress.com

Published by States Academic Press
109 South 5th Street,
Brooklyn, NY 11249, USA
www.statesacademicpress.com

Optics: Theory and Applications
Carl MacKenzie

International Standard Book Number: 978-1-63989-397-3 (Hardback)

Cataloging-in-Publication Data

Optics : theory and applications / Carl MacKenzie.
p. cm.
Includes bibliographical references and index.
ISBN 978-1-63989-397-3
1. Optics. 2. Physics. I. MacKenzie, Carl.
QC355.3 .O58 2022
535--dc23

Contents

Preface

The branch of physics which attempts to study the properties and behaviour of light is known as optics. It also deals with the interactions of light waves with matter and construction of instruments that are used to detect them. Electromagnetic waves such as ultraviolet light, infrared light, microwaves and radio waves are also studied under this discipline. The subject of optics can be further divided into classical optics, physical optics and modern optics. Within classical optics, light is considered to be an electromagnetic wave which travels in a straight line. Physical optics deals with the wave nature of light and studies phenomena such as superposition, dispersion and interference. Modern optics deals with study of new devices and technologies like lasers, photomultipliers, image sensors, charge coupled devices and quantum electronics. This book is compiled in such a manner, that it will provide in-depth knowledge about the theory and applications of optics. Some of the diverse topics covered herein address the varied branches that fall under this category. For all those who are interested in optics, this book can prove to be an essential guide.

A short introduction to every chapter is written below to provide an overview of the content of the book:

Chapter 1 - The branch of physics which deals with the study of behavior and properties of light is known as optics. It also includes the interaction of light with matter and the instruments associated with the detection of light. This is an introductory chapter which will introduce briefly all the significant aspects of optics and optical systems; **Chapter 2 -** Physical optics is the study of various phenomena of light where it behaves like electromagnetic wave and not like a particle. Some of the phenomena studied under physical optics are superposition, interference, diffraction, dispersion and polarization. This chapter discusses in detail the theories and phenomena related to physical optics; **Chapter 3 -** Geometrical optics is the branch of optics that studies the phenomena where light behaves as rays. Various phenomena such as reflection, specular reflection, diffuse reflection, signal reflection, total internal reflection, reflection phase change, refraction and refractive index. This chapter has been carefully written to provide an easy understanding of the varied facets of geometrical optics; **Chapter 4 -** There are various applications of optics such as microscopes, optical telescopes, optical periscopes, cameras and optical computing, etc. Optics also has applications in the field of healthcare and life sciences, astronomy and space and industrial manufacturing. The diverse applications of optics in the current scenario have been thoroughly discussed in this chapter; **Chapter 5 -** The means of communication which uses light for transmitting information is known as optical communication. The systems which are used for transmitting information optically through fibers by converting the electrical signal into light pulses are known as optical communication systems. The topics elaborated in this chapter will help in gaining a better perspective about optical communication systems, optical fibers and their applications.

I extend my sincere thanks to the publisher for considering me worthy of this task. Finally, I thank my family for being a source of support and help.

Carl MacKenzie

1

Introduction to Optics

The branch of physics which deals with the study of behavior and properties of light is known as optics. It also includes the interaction of light with matter and the instruments associated with the detection of light. This is an introductory chapter which will introduce briefly all the significant aspects of optics and optical systems.

Like all the different types of light, the spectrum of visible light is absorbed and emitted in the form of tiny packets of energy called photons. These photons have both the properties of a wave as well as a particle. Hence this type of property is called wave-particle duality and the study of light in the area of physics is known as Optics.

Optics is a branch of physics that deals with the determination of behavior and the properties of light, along with its interactions with the matter and also with the instruments which are used to detect it. Optics, in a simple manner, is used to describe the behavior of visible light, infrared light, and the ultraviolet. Imaging is done with the help of a system called an image forming an optical system.

Light and its Optical Properties

Light is a form of energy that is in the form of an electromagnetic wave and is almost everywhere around us. The visible light has wavelengths measuring between 400–700 nanometres. The Sun is the primary source of light by which plants utilize this to produce their energy.

In physics, the term light also refers to electromagnetic radiation of different kinds of wavelengths, whether it is visible to the naked eye or not. Hence, by this, the gamma rays, microwaves, X-rays, and radio waves are also types of light.

Light exhibits various properties which are given below:

- Reflection: Reflection is one of the primary properties of light. Reflection is nothing but what you see the images in the mirrors. Reflection is defined as the change in direction of light at an interface in-between two different media so that the wave-front returns into a medium from which it was originated. The typical examples for reflection of light include sound waves and water waves.
- Speed of light: The rate at which the light travels in free space is called the Speed of light. For example, the light travels 30% slower in the water when compared to vacuum.
- Refraction: The bending of light when it passes from one medium to another is called Refraction. This property of refraction is used in a number of devices like microscopes, magnifying lenses, corrective lenses, and so on. In this property, when the light is transmitted through a medium, polarization of electrons takes place which in-turn reduces the speed of light, thus changing the direction of light.

- Total internal reflection: When a beam of light strikes the water, a part of the light is reflected, and some part of the light is refracted. This phenomenon is called as Total internal reflection.
- Dispersion: It is a property of light, where the white light splits into its constituent colors. Dispersion can be observed in the form of a prism. The other properties of light include diffraction and interference. So, what you observe when, you look out at the beautiful scenario? Whether the light gets reflected, dispersed, refracted, internally reflected, or diffracted.

Applications of Optics

The properties of optics are applied in various fields of Physics:

- The refraction phenomenon is applied in the case of lenses (Convex and concave) for the purpose of forming an image of the object.
- Geometrical optics is used in studies of how the images form in an optical system.
- In medical applications, it is used in the optical diagnosis of the mysteries of the human body.
- It is used in the therapeutical and surgeries of the human tissues.

Types of Optics

Non-Linear Optics

When an electromagnetic wave propagates through a material, the particles of the medium are displaced from their equilibrium positions. Positively charged particles move in the direction of the field, while negatively charged particles move in the direction opposite to the direction of the electric field. This relative displacement of positively and negatively charged particles creates dipole moments. The dipole moment per unit volume describes the induced polarization P of the medium.

In a linear material the relative displacement of positively and negatively charged particles is proportional to the instantaneous magnitude of the electric field. But all real materials, when exposed to a high enough light intensity show a nonlinear response. In return the medium modifies the electromagnetic field in a nonlinear way.

When the magnitude of the external field is small, the magnitude of the electric polarization is approximately linearly proportional to the electric field E. In an isotropic material we write,

$$P = \varepsilon_o \chi_e E.$$

In an anisotropic material P and E can have a different direction. When the magnitude of external electric field is large, the induced polarization has a nonlinear dependence on the electric field and can be expressed as a power series with respect to the electric field. In an isotropic material we have:

$$P = \sum_n \varepsilon_o \chi^{(n)} E^n.$$

In an anisotropic material P and E^n can have a different direction.

$\chi^{(1)}$ is the linear susceptibility linear absorption and refraction. $\chi^{(2)}$ is the first non-linear term. It only is nonzero in anisotropic materials. $\chi^{(3)}$ is the second non-linear term, which can be nonzero in all materials.

Electric fields associated with light from ordinary non-coherent sources are usually too small to trigger an observable nonlinear response. But the magnitude of electric fields from pulsed laser beams focused onto a small spot can approach the magnitude of typical internal fields in crystals. In general it takes fields greater than of 10^5 V/m to observe nonlinear optical phenomena. These optical fields are easily generated by lasers. Equally important, the coherence of the laser light makes it possible to observe many nonlinear phenomena. When the molecules in the material respond coherently to the laser light, even the weakest nonlinear effects can be detected.

Assume E is proportional to cos(ωt). Then P is is proportional to cos(ωt) in a linear material. All charges oscillate with angular frequency ω and produce electromagnetic wave with angular frequency ω. In nonlinear materials P also contains non-negligible terms that are proportional to E^2 and therefore to $\cos^2(\omega t) = (1/2)(1 + \cos(2\omega t))$, and maybe terms proportional to E^3 and therefore to $\cos^3(\omega t) = (1/4)(\cos(3\omega t) + 3\cos(\omega t))$. The displacement of the charges is a superposition of a constant displacement and oscillations with angular frequencies ω, 2ω, and maybe 3ω. The material generates EM wave with twice or three times the frequency of the incident light and also a constant electric field.

The nonlinear index of refraction is defined as a variation of the index of refraction of the material proportional to the intensity of the light propagating through the medium. Let us consider an optical wave propagating through an isotropic material with inversion symmetry ($\chi^{(2)} = 0$). The linear susceptibility $\chi^{(1)}$ is related to the ordinary index of refraction in the linear regime.

$$n_o = (1 + \chi^{(1)})^{1/2}.$$

In the nonlinear regime the magnitude of the polarization can be expressed as:

$$P = \varepsilon_o (\chi^{(1)}E + \chi^{(3)} E^3).$$

This yield,

$$D = \varepsilon_o E + P = \varepsilon_o(1 + \chi^{(1)} + \chi^{(3)} E^2)E = \varepsilon E.$$

The index of refraction is,

$$n = (\varepsilon/\varepsilon_o)^{1/2} = (1 + \chi^{(1)} + \chi^{(3)} E^2)^{1/2} \sim n_o(1 + (\chi^{(3)} E^2)/2n_o^2) = n_o + n_2 I,$$

where I is the intensity of the optical wave, $I \propto E^2$, and n_2 represents the nonlinear index of refraction, which depends on the intensity of the applied field. The nonlinear index of refraction is responsible for self-focusing and soliton propagation in fibers.

The electro-optic effect is a nonlinear effect that mixes a static field with the field of and electromagnetic wave. If an isotropic dielectric is placed in an electric field E and a beam of light is passed through the sample perpendicular to the field then the material displays induced birefringence proportional to E^2. The ordinary and the extraordinary ray propagate in the same direction perpendicular to E, but the index of refraction is different for the ordinary and extraordinary ray. $\Delta n = n_o - n_e = \lambda K E^2$, where K is the Kerr coefficient. The Kerr effect is a third order effect, and it is sometimes referred to as the quadratic electro-optic effect. It can be generated in materials with any molecular orientation. It can be used to construct wave plates.

Certain crystals exhibit a linear electro-optic effect. Birefringence occurs when the material is placed in an electric field E and a beam of light passes through the material parallel to the field. The ordinary and the extraordinary ray propagate in the same direction parallel to E, but the index of refraction is different for the ordinary and extraordinary ray. The induced birefringence is proportional to E. $\Delta n = n_o K_p E$ where K_p is the Pockels coefficient, and n_o is the index of refraction of the material with no field applied. The linear electro-optic effect (LEO) or Pockels effect is a 2nd-order nonlinear effect. It does not occur in isotropic media or media with inversion symmetry. The Pockels effect can also be used to construct wave plates.

Both the Pockels and the Kerr effects can be used to construct very fast optical shutters (10^{-10} s) by placing the material between crossed polarizers. When the retardation is $\lambda/2$, the device will be transparent. Pockels cells typically require 5 to 10 times lower voltages than equivalent Kerr cells.

Second harmonic generation is a second order nonlinear process ($\omega + \omega$ --> 2ω). Parametric up-conversion in a crystal can be used to convert a signal from a low frequency ω_1 to a high frequency ω_3 by mixing it with a strong laser beam of frequency ω_2 ($\omega_1 + \omega_2$ --> ω_3). Parametric amplification involves an input signal at at ω_1 together with an intense pump beam at ω_2, $\omega_2 > \omega_1$. A photon of the pump beam with frequency ω_2 interacts with a photon with frequency ω_1 and splits into two photons, one with frequency ($\omega_2 - \omega_1$) and one with frequency ω_1. The coupling is provided by the non-linearity of the crystal. The wave at ω_1 is amplified. This is accompanied by the generation of an idler beam at $\omega_3 = \omega_2 - \omega_1$. The idler beam then interacts with the pump beam to produce additional amounts of signal and idler light. A photon of the pump beam with frequency ω_2 interacts with a photon of frequency $\omega_3 = \omega_2 - \omega_1$ and splits into two photons, one with frequency $\omega_2 - \omega_3 = \omega_1$ and one with frequency ω_3.

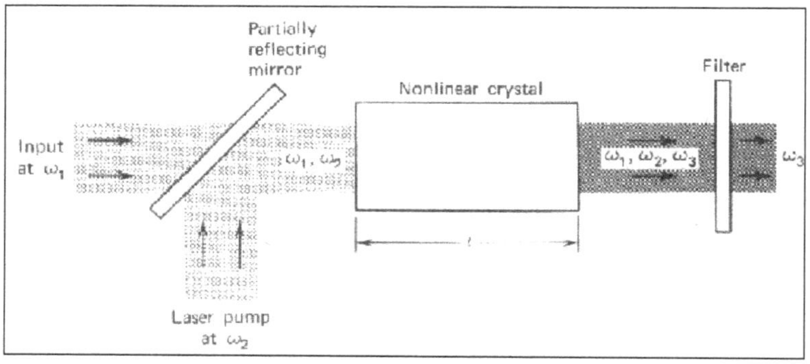

In most materials the index of refraction depends on the wavelength of the EM wave, EM waves with different angular frequencies ω propagate with slightly different speeds. Consider, for example, second harmonic generation. Light with frequency 2ω generated at some point P_1 in the medium will not be in phase with light with frequency 2ω generated at some other point P_2. As a result, the two waves from P_1 and P_2 will interfere destructively at some point P_3. The problem of destructive interference can be overcome by letting the fundamental be the ordinary ray and the second harmonic be the extraordinary ray in an anisotropic crystal. When the direction of propagation of the fundamental is chosen so that the index of refraction of the fundamental and second harmonic is the same, both waves will stay in phase. The crystal then is said to be phase-matched or index-matched. Phase matching can also be achieved in some crystals by varying their temperature. The extraordinary index is in general more temperature-dependent than the ordinary index. By varying the temperature one can adjust the birefringence of the crystal until phase matching is obtained.

Other Induced Optical Effects

Photo-elasticity

An isotropic material can become birefringent when placed under stress. Under compression it becomes a negative uniaxial crystal and under tension it becomes a positive uniaxial crystal. The direction of the stress defines the direction of the optical axis. The induced birefringence is proportional to the stress. Photo-elasticity can be used to study stress patterns in complex objects, for example bridges, by building a transparent scale model of the object.

Faraday Rotation in Solids

If an isotropic dielectric is placed in a magnetic field B and a beam of linearly polarized light is passed through the sample in the direction of the field then a rotation of the plane of polarization will occur. The angle θ through which the direction of polarization is rotated is proportional to B and the thickness of the material. The proportionality constant is called the Verdet constant. The direction of the rotation depends on whether light is traveling parallel or anti-parallel to B.

Electron and Ion Optics

The branch of science that studies the behavior of electron and ion beams in a vacuum under the influence of electric and magnetic fields. Inasmuch as the study of electron beams began before the study of ion beams and the former are used much more extensively than the latter, the narrower term "electron optics" is widely used. Electron and ion optics deals mainly with problems of the formation, focusing, and deflection of charged particle beams and the production, by means of such beams, of optical images on a luminescent screen or photographic film. Such images are commonly called electron-optical and ion-optical images. The development of electron and ion optics has largely been in response to the requirements of electronics technology.

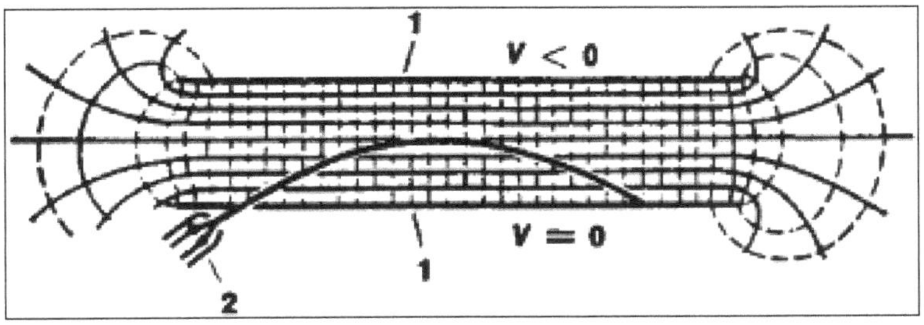

Deflection of an electron beam in the uniform field of a parallel-plate capacitor: (1) capacitor plates, (2) electron emitter producing the electron beam. The lines of force are represented by broken lines, and the intersections of the equipotential surfaces and the plane of the figure are represented by solid lines. The field potential V increases from top to bottom.

The rise of electron and ion optics was associated with the creation of the electron-beam tube in the late 19th century. In the first oscilloscope electron-beam tube, constructed by C. F. Braun in 1897, the electron beam was deflected by a magnetic field. Deflection by means of an electrostatic field was accomplished by J. J. Thomson; in his experiments to determine the ratio of the electron's charge to its mass, Thomson passed a beam through a parallel-plate capacitor located inside an electron-beam tube. In 1899 a coil of insulated wire through which an electric current flowed was used to focus an electron beam in an electron-beam tube. It was not until 1926, however, that the German scientist H. Busch theoretically considered the motion of charged particles in the magnetic field of such a coil; Busch demonstrated that the coil was suitable for producing accurate electron-optical images and thus constituted an electron lens.

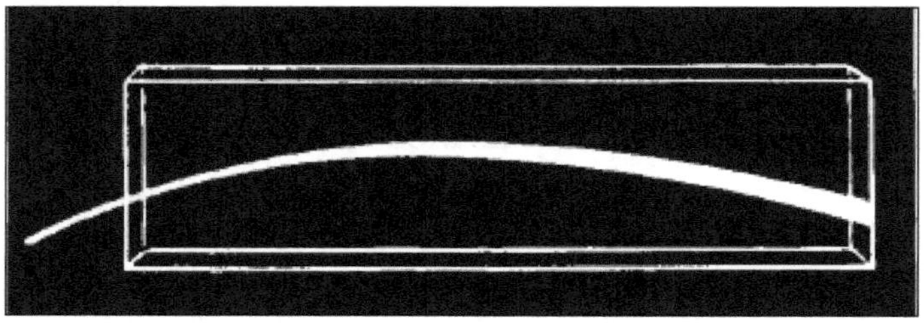

Deflection of a beam of light in an inhomogeneous medium. The refractive index n increases from top to bottom. The optical medium is an inhomogeneous mixture of carbon disulfide (n = 1.63) and benzene (n = 1.5) in a transparent vessel. The percent of carbon disulfide in the mixture increases from 0 to 100 percent from the top layers of the liquid to the bottom layers.

The subsequent development of electron lenses (both magnetic and electrostatic) opened the way to the creation of the electron microscope, image converter, and various other devices for the production of accurate electron-optical images of objects that either emit electrons or in some way affect electron beams. The construction of specialized electron-beam tubes for use in television and radar equipment and for data recording, storage, and reproduction led to the further development of the branches of electron and ion optics associated with the control of charged particle beams. The development of equipment for the analysis of electron and ion flows, such as beta spectrometers, mass spectrometers, and other analytical devices, significantly influenced the development of electron and ion optics.

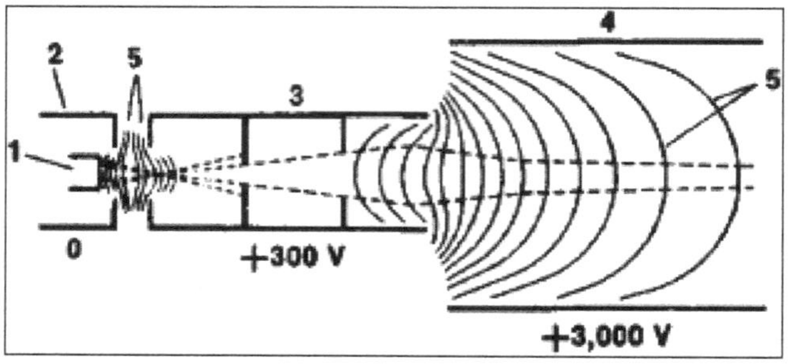

Electron-optical system with rotational symmetry, designed to produce an electron beam (an electron gun): (1) indirectly heated cathode, (2) focusing electrode, (3) first anode, (4) second anode, (5) intersections of equipotential surfaces of the electrostatic field and the plane of the figure. Broken lines indicate the beam contours. Shown beside the electrodes are their potentials with respect to the cathode, whose potential is assumed equal to zero. Electrodes (1), (2), and (3) form a cathode lens, and electrodes (3) and (4) form an immersion lens.

Electron and ion optics generally does not consider the problems that arise in microwave technology, and only rarely treats processes occurring in electron tubes, particle accelerators, and other equipment and devices whose specific characteristics place them outside the branch's primary field of interest.

The solution of most of the problems of electron and ion optics requires consideration only of the motion of charged particles within the framework of classical mechanics, since the wave nature of particles essentially is not manifested in such problems. In such an approximation electron and ion optics is called geometrical electron and ion optics because of the consistent analogy with the geometrical optics of light rays; the analogy is grounded in the similarity between the behavior of charged particle beams in electric and magnetic fields and that of light beams in inhomogeneous optical media. The similarity is apparent from a comparison of figures. The analogy is underlain by the more general analogy between classical mechanics and the geometrical optics of light that was established by W. R. Hamilton, who proved in 1834 that the general equation of mechanics (the Hamilton-Jacobi equation) is similar in form to the optical eikonal equation. As in the geometrical optics of light, the concept of the refractive index is introduced in geometrical electron and ion optics, and the eikonal method is often used in calculating image errors (aberrations), most of which are analogous to the aberrations of optical systems. When the approximation of geometrical electron and ion optics is inadequate, for example, in the investigation of the resolution of an electron microscope, the techniques of quantum mechanics are used.

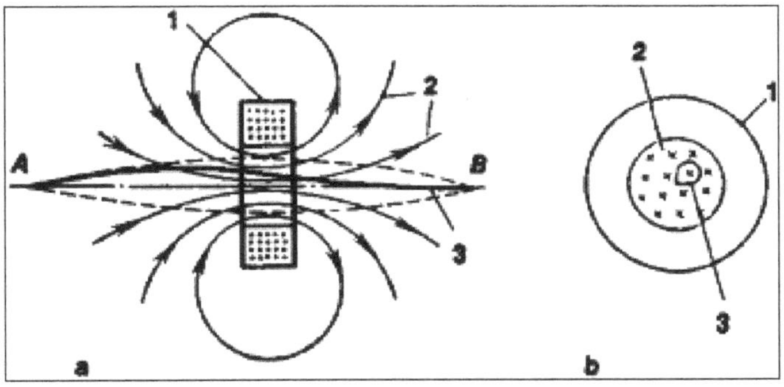

Magnetic lens in the form of a toroidal coil: (a) side view, (b) front view; (1) coil, (2) magnetic lines of force, (3) electron trajectory. Broken lines indicate the contours of the electron beam, which emerges from point A (the object) and is focused at point B.

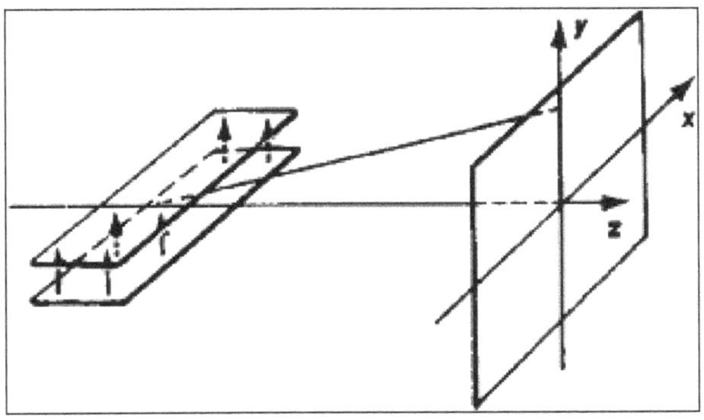

Deflection of a beam of positively charged particles in the field of a parallel-plate electrostatic capacitor. The arrows indicate the direction of the electric field inside the capacitor.

Electric and magnetic fields that have rotational (circular) symmetry with respect to the optical axis of a system are used extensively in electron-optical devices. Electron lenses and electron mirrors with such fields are said to be axially symmetric. Electric fields with rotational symmetry are generated by electrodes in the shape of cylinders, cups, and diaphragms with round apertures. Electromagnets (occasionally, permanent magnets) with pole pieces in the form of bodies of revolution or toroidal coils with a winding of insulated wire, through which an electric current is passed, are used to generate axially symmetric magnetic field.

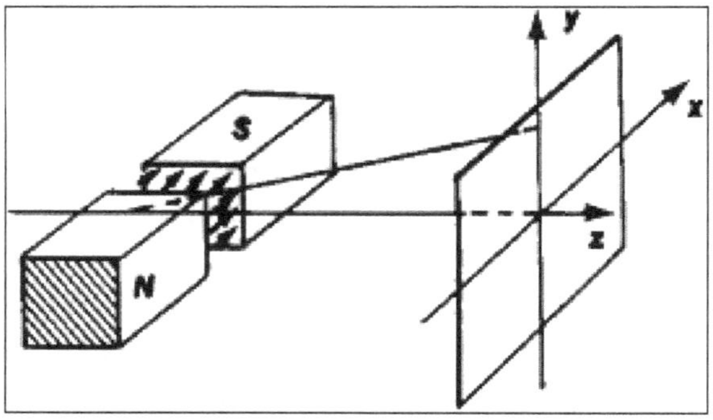

Deflection of a beam of positively charged particles by a transverse magnetic field: (*N*) and (*S*) magnetic poles. The arrows indicate the direction of the magnetic field in the gap between the poles.

Axially symmetric lenses and mirrors produce accurate electron-optical images if the charged particles travel sufficiently close to the field axis of symmetry and if their initial velocities differ little from one another. If these conditions are not fulfilled, the image errors become significant. When the object and image lie outside the field, axially symmetric electron lenses always constitute collecting lenses. In axially symmetric electron lenses, as in normal optical lenses with spherical surfaces, the image can be only direct or inverted, but in magnetic electron lenses the image additionally may be rotated by some angle.

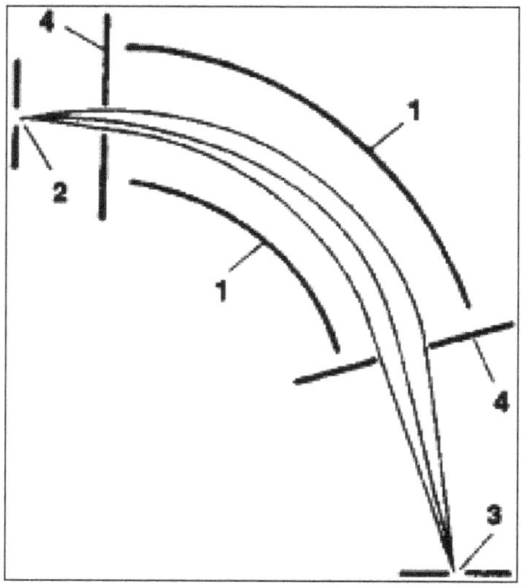

Deflection and focusing of a beam of charged particles by a sectional tubular capacitor: (1) capacitor electrodes, (2) exit slit of the source of charged particles, (3) entrance slit of the particle detector, (4) beam-limiting apertures. The electrodes have the shape of sections of circular cylinders. The source slit acts as the object. The divergent particle beam emerges from the source slit with a specific energy and is focused to form a linear image of the source slit; the image is perpendicular to the plane of the figure and aligned with the detector slit.

The electron-optical properties of a field with rotational symmetry are determined by the location of the cardinal points, which are analogous to the cardinal points of axially symmetric optical imaging systems: two foci, two principal points, and two nodal points. The image is constructed according to the rules of geometrical light optics. Axially symmetric electrostatic fields exhibit the same five types of third-order geometrical aberrations as do centered optical systems with spherical surfaces: spherical aberration, astigmatism, curvature of image field, distortion, and coma. Magnetic fields exhibit three additional aberrations: anisotropic distortion, anisotropic astigmatism, and anisotropic coma.

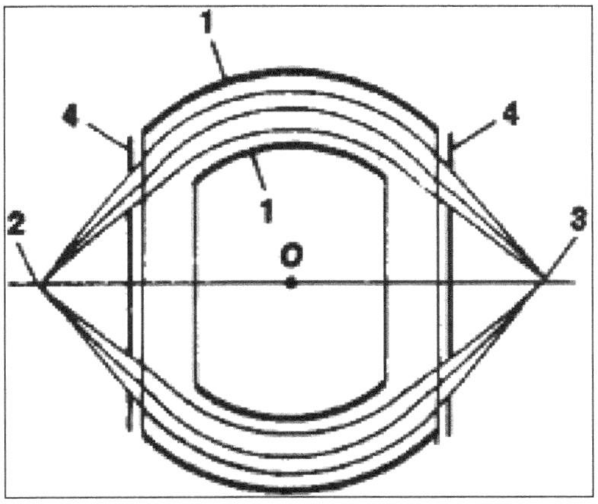

Spherical capacitor: (1) capacitor electrodes, (2) point object, (3) image of object, (4) beam-limiting apertures. The electrodes have the shape of sections of two concentric spheres. The image lies on the straight line passing through the source and the center O of the spheres.

Furthermore, there are three types of chromatic aberration (two in electrostatic fields), caused by some unavoidable energy spread of the particles entering the field. In general, the aberrations of fields with rotational symmetry exceed in magnitude, under comparable conditions, the aberrations of centered optical systems; that is, electron lenses and electron mirrors are substantially inferior in quality to optical systems. The compensation for or reduction of aberrations is one of the basic problems in theoretical electron and ion optics.

Other types of electron lenses and mirrors have fields that exhibit different types of symmetry. They form images of point objects as line segments, but sometimes are capable of providing stigmatic focusing (point-to-point focusing) as well. Cylindrical electrostatic and magnetic lenses and mirrors produce line images of point objects. The fields in such electron lenses are "two-dimensional" (their intensities are described by functions of just two Cartesian coordinates) and are symmetric with respect to some central plane near which the charged particles move. Precise focusing in just one direction is needed in many analytical electric vacuum devices. In such cases it is expedient to use transaxial electrostatic electron lenses or transaxial electron mirrors, whose aberrations in the central plane are very small (comparable to the aberrations of optical lenses). Quadrupole electron lenses (electrical and magnetic) are used for work with high-energy charged particle beams.

Electron-optical devices with electric or magnetic fields directed across the beam are used to deflect charged particle beams. The parallel-plate capacitor represents the simplest electrical deflecting element. In order to reduce the deflecting voltage, systems with electrodes of more complex shape may be used in electron-beam tubes. The magnetic fields used to deflect beams may be generated by electromagnets or by conductors through which a current flows.

In analytical devices that use deflecting electric and magnetic fields to separate (resolve) charged particles with respect to energy and mass, the forms of the fields used vary widely. Extensive use is also made of the fields' ability to focus beams. The electric fields are usually formed by various capacitors: parallel-plate, tubular, and spherical types. Of the magnetic fields, uniform fields and sectoral fields are often used. The boundaries of sectoral magnetic fields may be bent in order to improve the quality of focusing, and nonuniform magnetic fields whose strength varies according to a specific law may also be used for the same purpose.

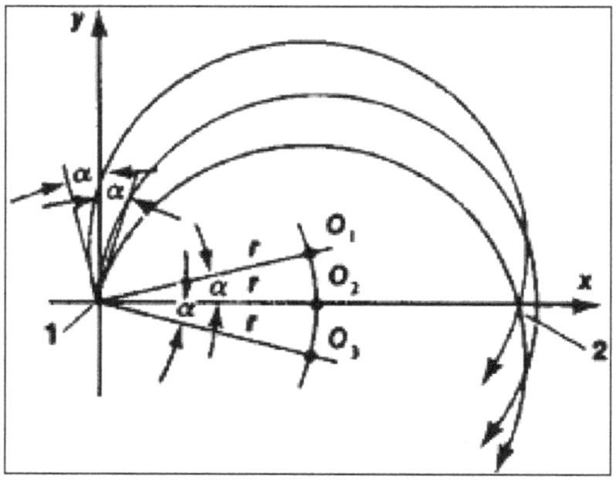

Deflection and focusing of a beam of charged particles by a uniform magnetic field: (1) object, (2) image; (O,), (O_1), and (O_1) centers of the circular trajectories of the particles. The charged particles emitted by a linear object (slit) within the small angle 2α initially diverge and are then focused, describing semicircles of radius r, which is the same for all particles of identical mass and energy. The focused particles form an image of the object in the shape of a strip with a width of $r\alpha^2$. The linear object and strip image are situated parallel to the magnetic lines of force, which are perpendicular to the plane of the figures.

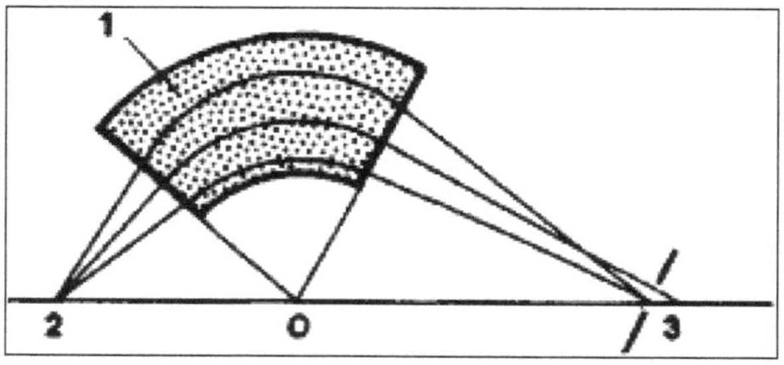

Deflection and focusing of a beam of charged particles by a sectoral magnetic field: (1) magnetic field, (2) object (source slit), (3) image. The magnetic lines of force are perpendicular to the plane of the figure. The image lies on the line connecting the object with the vertex O of the sector. The width of the image is of the same order as in a uniform magnetic field.

The deflecting electric and magnetic devices discussed above, which are sometimes called electron (ion) prisms, differ from optical prisms in that they not only deflect but also focus charged particle beams. As a result of focusing, parallel beams that enter the fields of such devices cease to be parallel after deflection. However, the creation of high-quality analytical electron and ion devices, constructed on an exact analogy with optical prism spectrometers, requires electron (ion) prisms that also preserve beam parallelism. Telescopic systems of electron lenses may be used for this function. By adding two electron lenses to an electron prism, with one (the collimator lens) at the inlet and the other (the focusing lens) at the exit, it is possible to construct an analytical device that combines high resolution and high electron-optical transmission.

Fiber Optics

Fiber optics, which is the science of light transmission through very fine glass or plastic fibers, continues to be used in more and more applications due to its inherent advantages over copper conductors.

Advantages

Fiber optics has many advantages over copper wire including:

- Increased bandwidth: The high signal bandwidth of optical fibers provides significantly greater information carrying capacity. Typical bandwidths for multimode (MM) fibers are between 200 and 600MHz-km and >10GHz-km for single mode (SM) fibers. Typical values for electrical conductors are 10 to 25MHz-km.
- Electromagnetic/Radio Frequency Interference Immunity: Optical fibers are immune to electromagnetic interference and emit no radiation.
- Decreased cost, size and weight: Compared to copper conductors of equivalent signal carrying capacity, fiber optic cables are easier to install, require less duct space, weigh 10 to 15 times less and cost less than copper.
- Lower loss: Optical fiber has lower attenuation (loss of signal intensity) than copper conductors, allowing longer cable runs and fewer repeaters.
- No sparks or shorts: Fiber optics do not emit sparks or cause short circuits, which is important in explosive gas or flammable environments.
- Security: Since fiber optic systems do not emit RF signals, they are difficult to tap into without being detected.
- Grounding: Fiber optic cables do not have any metal conductors; consequently, they do not pose the shock hazards inherent in copper cables.
- Electrical Isolation: Fiber optics allow transmission between two points without regard to the electrical potential between them.

Table: Advantages of Fiber Optics over Copper.

	Coaxial Cable	Fiber Optic Cable (MM)	Fiber Optic Cable (SM)
Representative distance bandwidth products	100 MHz km	500 MHz km	100,000+ MHz km
Attenuation/km @ 1 GHz	>45 dB	1 dB	0.2 dB
Cable cost ($/m)	$$$$$$$$$	$	$
Cable diameter (in.)	1	1/8	1/8
Data security	Low	Excellent	Excellent
EMI immunity	OK	Excellent	Excellent

Fiber Optic Link Components

In order to comprehend how fiber optic applications work, it is important to understand the components of a fiber optic link. Simplistically, there are four main components in a fiber optic link.

- Optical Transmitter.
- Optical Fiber/Cable.
- Connectors.
- Optical Receiver.

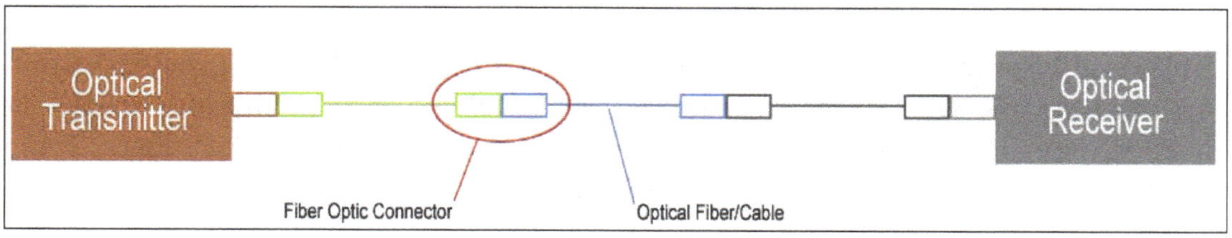

Simple Fiber Optic Link.

Transmitter

The transmitter converts the electrical signals to optical. A transmitter contains a light source such as a Light Emitting Diode (LED) or a Laser (Light Amplification by Stimulated Emission of Radiation) diode, or a Vertical Cavity Surface Emitting Laser (VCSEL).

- LED: Is used in multimode applications and has the largest spectral width that carries the least amount of bandwidth.
- VCSEL: Is also used in multimode applications with a narrower spectral width that can carry more bandwidth than the LED.
- LASER: Has the smallest spectral width, carries the most bandwidth, and is used in single-mode applications.

These sources produce light at certain wavelengths depending upon the materials from which they are made. Most fiber optic sources use wavelengths in the infrared band, specifically 850nm (1nm=10-9m), 1300nm and 1550nm. For reference, visible light operates in the 400-700nm range.

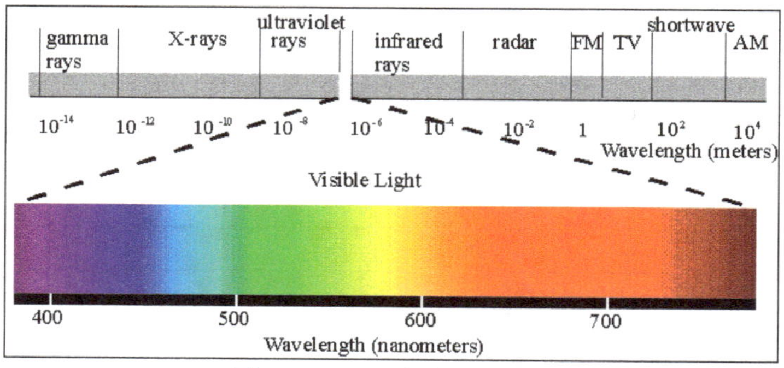

Electromagnetic Spectrum.

Optical Fiber/Cable

An optical fiber is made of 3 concentric layers:

- Core: This central section, made of silica or doped silica, is the light transmitting region of the fiber.
- Cladding: This is the first layer around the core. It is also made of silica, but not the same composition as the core. This creates an optical waveguide which confines the light in the core by total internal reflection at the core-cladding interface.
- Coating: The coating is the first non-optical layer around the cladding. The coating typically consists of one or more layers of polymer that protect the silica structure against physical or environmental damage. The coating is stripped off when the fiber is connectorized or fusion spliced.

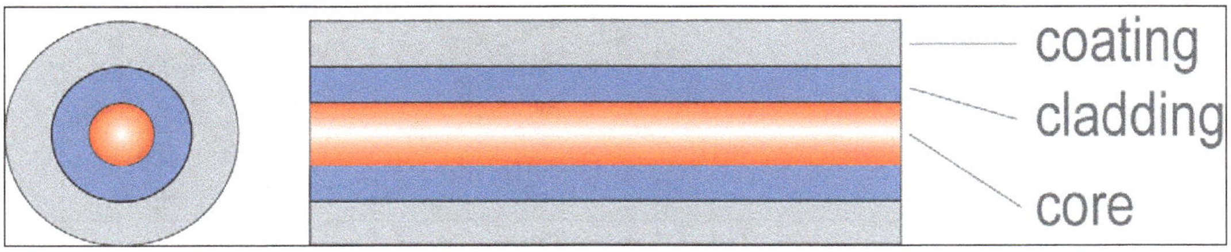

Optical Fiber Construction.

- Buffer (not pictured): The buffer is an important feature of the fiber. It is 900 microns and helps protect the fiber from breaking during installation and termination and is located outside of the coating.

The light is "guided" down the core of the fiber by the optical "cladding" which has a lower refractive index (the ratio of the velocity of light in a vacuum to its velocity in a specified medium) that traps light in the core through "total internal reflection."

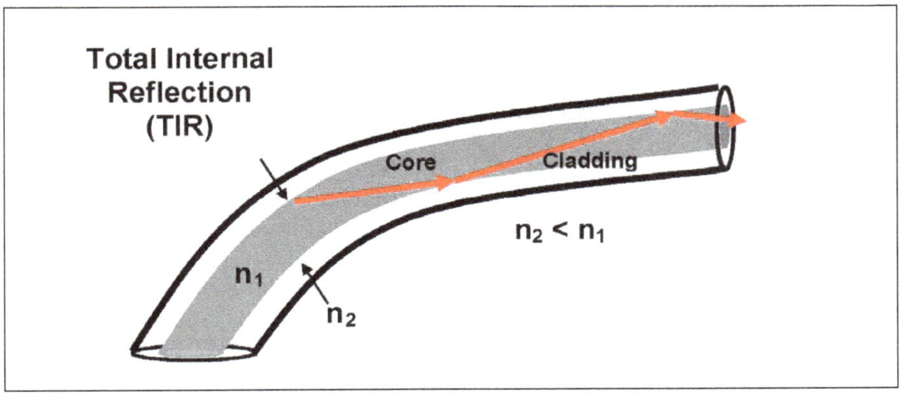

Diagram showing Total Internal Reflection.

In fiber optic communications, single mode and multimode fiber constructions are used depending on the application. In multimode fiber, light travels through the fiber following different light paths called "modes." In single mode fiber, only one mode is propagated "straight" through the fiber.

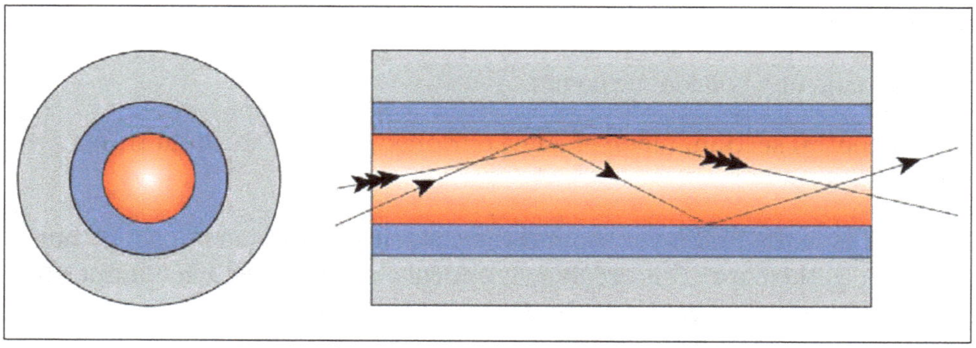

Multimode Fiber Light Propagation.

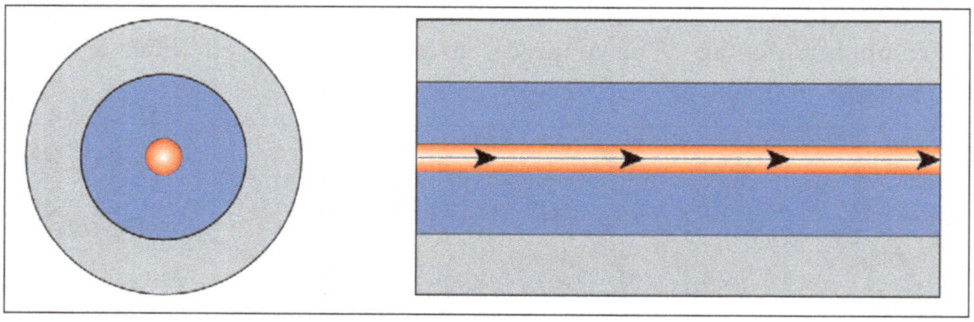

Single Mode Fiber Light Propagation.

Typical multimode fibers have a core diameter/cladding diameter ratio of 50 microns/125 microns (10-6 meters) and 62.5/125 (although 100/140 and other sizes are sometimes used depending on the application). Single mode fibers have a core/cladding ratio of 9/125 at wavelengths of 1300nm and 1550nm.

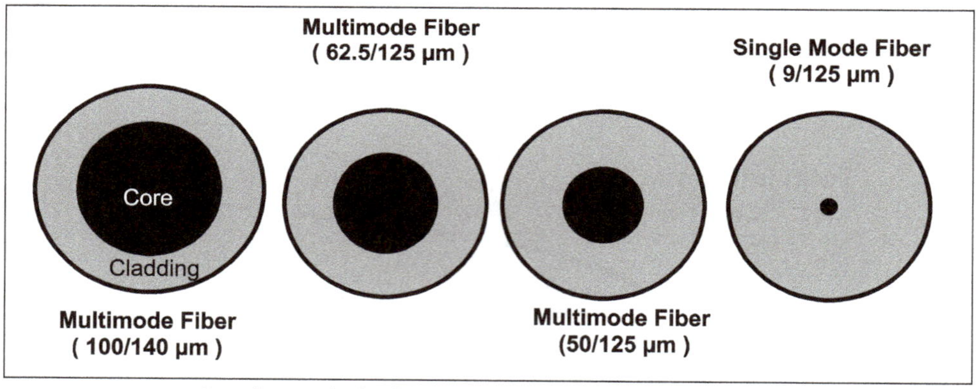

Popular Optical Fiber Core/Cladding Diameter Ratios.

Light is gradually attenuated when it travels through fiber. The attenuation value is expressed in dB/km (decibel per kilometer). Attenuation is a function of the wavelength (λ) of the light. Figure 8 shows the attenuation as a function of the wavelength.

The typical operating wavelengths are 850nm (nanometers) and 1300nm in multimode, and 1300nm or 1550nm in single mode. Note that there are natural "dips" in the attenuation graph at these wavelengths. For example, at an 850nm operating wavelength, there is 3dB attenuation after 1km propagation (according to the graph). 3dB of attenuation means that 50% of light has been lost.

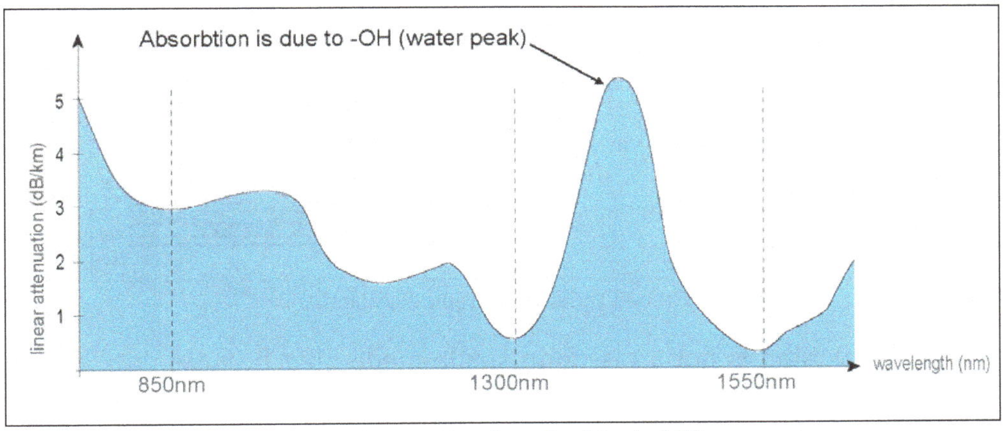

Attenuation vs. Wavelength of Optical Fiber.

Bandwidth is a measure of the data-carrying capacity of an optical fiber. It is expressed as the product of frequency and distance. For example, a fiber with a bandwidth of 500MHz-km (Megahertz kilometer) can transmit data at a rate of 500MHz along one kilometer of fiber. The bandwidth of single mode fibers is much higher than in multimode fibers. The main reason for the lower bandwidth in multimode fibers is modal dispersion.

In multimode fibers, information (ABC) is propagated in fiber according to N modes or paths, as if it were "duplicated" N times (for example, in the diagram, the mode 3 path is longer than the mode 2 path, which are both longer than the mode 1 path). If information is too close, there is a risk of overlapping ("smearing") the information, and then it will not be recoverable at the end of the fiber. It is necessary to space the data sufficiently to avoid overlap, i.e., to limit the bandwidth.

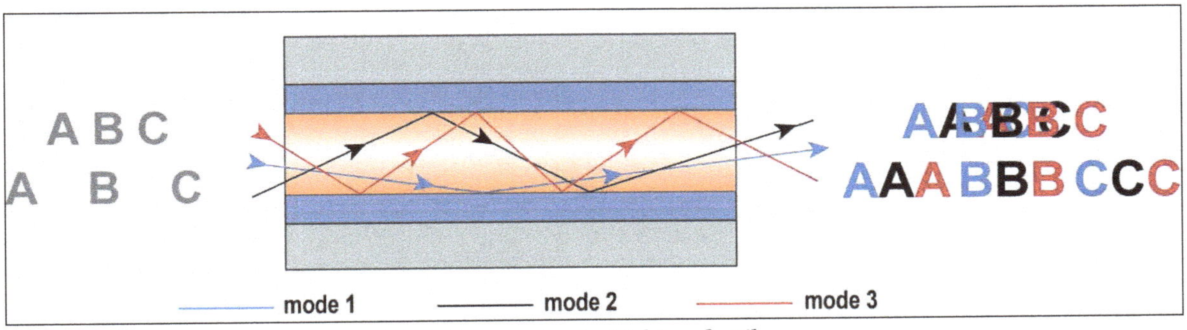

Modal Dispersion in Multimode Fibers.

Modal dispersion can be alleviated to a large extent by grading the index of refraction from the middle of the core to the cladding (graded index fiber), thereby equalizing the paths.

In a step index fiber, the index of refraction changes abruptly from the core to the cladding. To help reduce modal dispersion, fiber manufacturers created graded-index fiber. Graded-index fiber has an index of refraction which gradually increases as it progresses to the center of the core. Light travels slower as the index of refraction increases. Thus, a light path propagating directly down the center of the fiber has the shortest path but will arrive at the receiver at the same time as light that took a longer path due to the graded-index of the fiber.

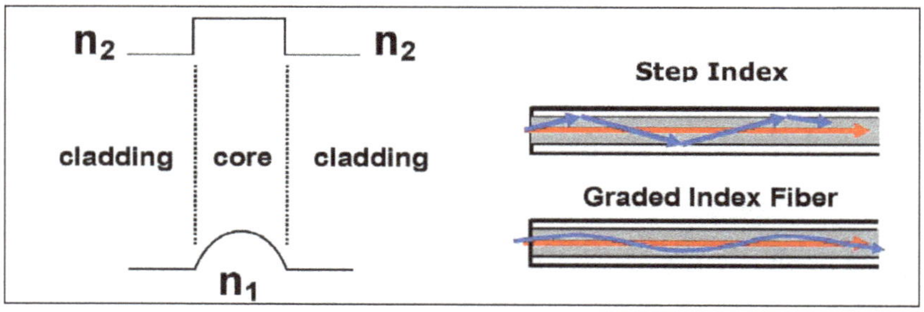

Graded Index in Multimode Fibers.

Of course, modal dispersion is not an issue in single mode fiber because only a single mode is propagated.

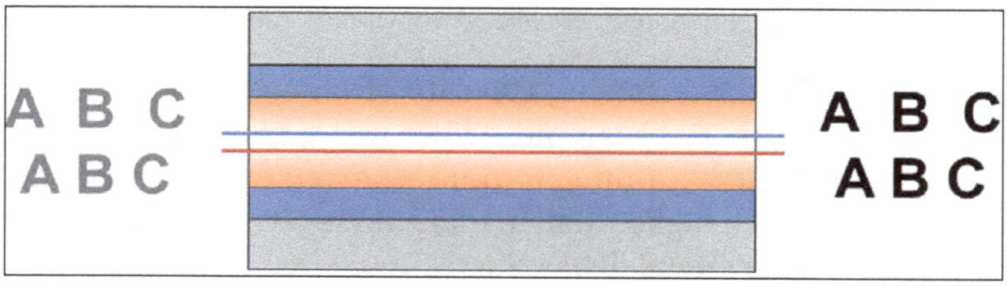

Single Mode Propagation.

Unfortunately, the optical fiber construction shown in Figure is fragile. Thus, for most applications, the fiber must be made into a cable. There are many ways to construct a cable (tight buffer, loose tube, gel filled, distribution, breakout, etc). However, in our single fiber cable example, the 250 micron coating is jacketed with a 900 micron buffer and built into a 3.0m outer sheath cable with aramid yarn as a strength member. As a typical example, Figure portrays a cable with multiple optical fibers.

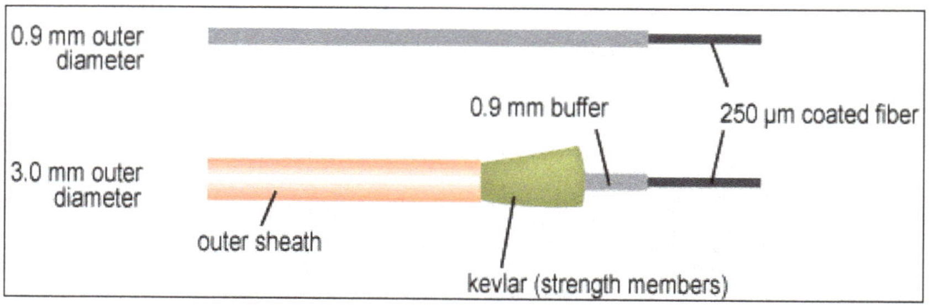

Construction of a Single Fiber Cable.

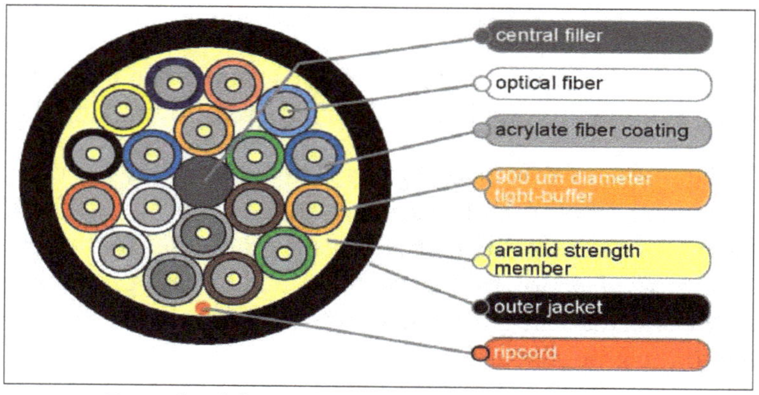

Example of the Construction of a Multi- Fiber Cable.

Connectivity

Fiber optic links require a method to connect the transmitter to the fiber optic cable and the fiber optic cable to the receiver. In general, there are two methods to link optical fibers together.

Fusion Splice

The first method is called a fusion splice. This operation consists of directly linking two fibers by welding with an electric arc or a fusion splicer. The advantages of this approach are that the linking method is fast and simple and there is very little insertion loss (the loss of light generated by a connection is called Insertion Loss [IL]). The disadvantages are that the link is relatively fragile, is permanent, and the initial cost (of the fusion splicer) is high.

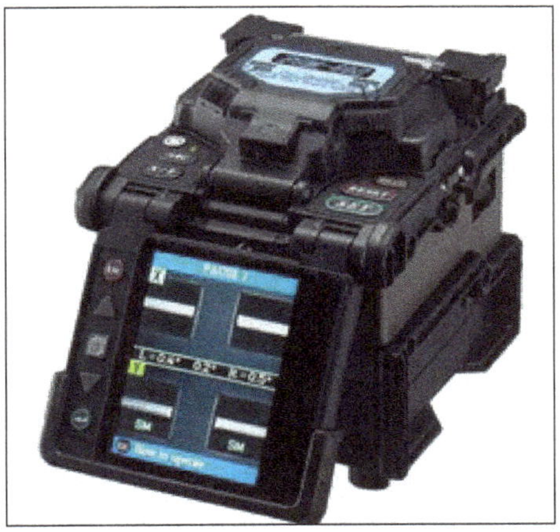

Fiber-optic Fusion Splicer.

Connectors

The second method involves the uses of fiber optic connectors. A connector terminates the optical fiber inside a ceramic ferrule, using epoxy to hold the fiber in place. The connectors can be mated and unmated at any time. The advantages of this approach are that the connection is robust, the connector can be chosen according to the application, and the connector can be connected and disconnected hundreds or even thousands of time without damaging the connectors. The disadvantages of this approach are that the connectorization takes longer than fusion splicing, requires special tools, and the insertion loss can be higher when compared with fusion splicing. There are two types of fiber optic connectors: physical contact and expanded beam.

Physical Contact Connectors

Physical contact connectors utilize fiber in a tightly toleranced ceramic ferrule. This allows easy handling of the fiber and protects it from damage. The principle of physical contact connectors involves the direct contact of polished fibers within two ceramic ferrules. The ferrules are aligned using a ceramic alignment sleeve. Insertion loss is a function of the alignment accuracy and the polish quality. There are springs behind the ferrule to ensure that the two ferrules are in constant contact even in high vibration and shock environments.

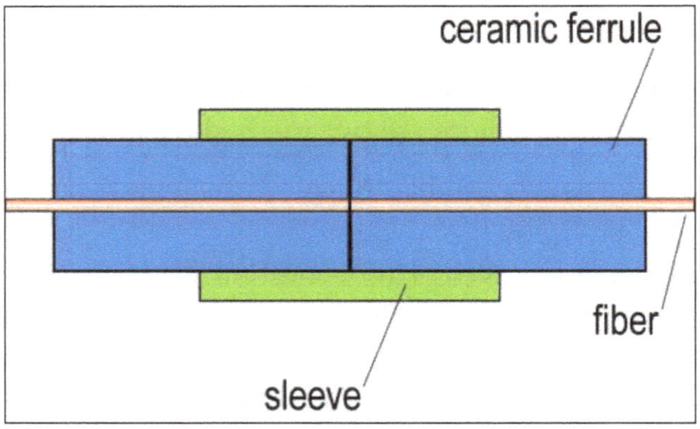

Physical Contact.

Optical System

An optical system consists of a succession of elements, which may include lenses, mirrors, light sources, detectors, projection screens, reflecting prisms, dispersing devices, filters and thin films, and fibre-optics bundles.

Lenses

All optical systems have an aperture stop somewhere in the system to limit the diameter of the beams of light passing through the system from an object point. By analogy with the human eye, this limiting aperture stop is called the iris of the system, its images in the object and image spaces being called the entrance pupil and exit pupil, respectively. In most photographic lenses the iris is inside the objective, and it is often adjustable in diameter to control the image illumination and the depth of field. In telescope and microscope systems the cylindrical mount of the objective lens is generally the limiting aperture or iris of the system; its image, formed behind the eyepiece where the observer's eye must be located to see the whole area being observed, called the field, is then the exit pupil.

The pupils of a lens system can be regarded as the common bases of oblique beams passing through the system from all points in an extended object. In most systems, however, the mounts of some of the lens elements cut into the oblique beams and prevent the beams from being perfectly circular, and the pupils are then not fully filled with light. This effect is known as vignetting and leads to a reduction in illumination in the outer parts of the field of view.

A common feature of many optical systems is a relay lens, which may be introduced to invert an image or to extend the length of the system, as in a military periscope. An example of the use of a relay lens is found in the common rifle sight shown in Figure. Here the front lens A is the objective, forming an inverted image of the target on the cross wire or reticle at B. The light then proceeds to the relay lens C, which forms a second image, now erect, at D. Beyond this image is the eyepiece E to render the light parallel so that the image may be seen sharply by the observer. Unfortunately, the oblique beam from the objective will usually miss the relay lens, and so a field lens must be inserted at or near the first image B to bend the oblique beams around and redirect them toward the relay lens. The power of the field lens is chosen so that it will form an image of the objective lens aperture on the relay lens aperture. The iris and entrance pupil of this system coincide at the objective; there is an internal pupil at the relay lens, and the exit pupil lies beyond the eyepiece as shown in figure.

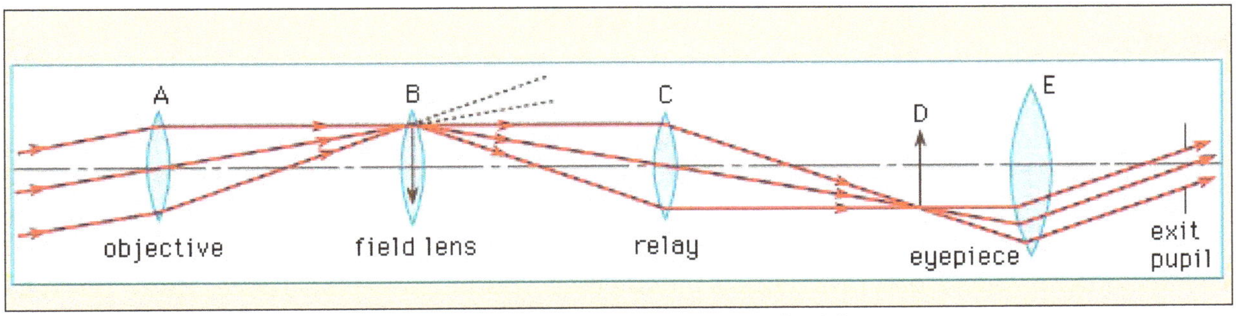

Operating principle of the telescopic rifle sight.

Mirrors

Mirrors are frequently used in optical systems. Plane mirrors may be employed to bend a beam of light in another direction, either for convenience or to yield an image reversed left for right if required. Curved mirrors, concave and convex, may be used in place of lenses as image-forming elements in reflecting telescopes. All of the world's largest telescopes and many small ones are of the reflecting type. Such telescopes use a concave mirror to produce the main image, a small secondary mirror often being added to magnify the image and to place it in a convenient position for observation or photography. Telescope mirrors are commonly made parabolic or hyperbolic in section to correct the aberrations of the image. Originally telescope mirrors were made from polished "speculum metal," an alloy of copper and tin, but in 1856 Justus von Liebig, a German chemist, invented a process for forming a mirror-like layer of silver on polished glass, which was applied to telescope mirrors by the German astronomer C.A. von Steinheil. Today most mirrors are made of glass, coated with either a chemically deposited silver layer or more often one made by depositing vaporized aluminum on the surface. The aluminum surface is as highly reflective as silver and does not tarnish as readily.

A large astronomical mirror presents many problems to the optical engineer, mainly because even a distortion of a few microns of the mirror under its own weight will cause an intolerable blurring of the image. Though many schemes for supporting a mirror without strain have been tried, including one to support it on a bag of compressed air, the problem of completely eliminating mirror distortion remains unsolved. A metal mirror, if well ribbed on the back, may be lighter than a glass mirror and therefore easier to handle, but most metals are slightly flexible and require just as careful support as glass mirrors. Since temperature changes can also cause serious distortion in a mirror, astronomers try to hold observatory temperatures as constant as possible.

Light Sources

Many types of optical instruments form images by natural light, but some, such as microscopes and projectors, require a source of artificial light. Tungsten filament lamps are the most common, but if a very bright source is required, a carbon or xenon arc is employed. For some applications, mercury or other gas discharge tubes are used; a laser beam is often employed in scientific applications. Laser light is brilliant, monochromatic, collimated (the rays are parallel), and coherent (the waves are all in step with each other), any or all of these properties being of value in particular cases.

Detectors

The image formed by an optical system is usually received by the eye, which is a remarkably adaptable and sensitive detector of radiation within the visible region of the electromagnetic spectrum. A photographic film, another widely used detector, has the advantage of yielding a permanent record of events. Since about 1925 many types of electrical detectors of radiation, both within the visible region and beyond it, have been developed. These include photoelectric cells of various kinds in which either a voltage or a resistance is modified by light falling on the device. Many new types of detectors are sensitive far into the infrared spectrum and are used to detect the heat radiated by a flame or other hot object. A number of image intensifiers and converters, particularly for X-ray or infrared radiation, which have appeared since World War II, embody a radiation detector at one end of a vacuum tube and an electron lens inside the tube to relay the image on to a phosphor screen at the other end. This arrangement produces a visible picture that may be observed by eye or photographed to make a permanent record.

Television camera tubes detect real images by electronic scanning, the picture on the viewing tube being a replica of the image in the original camera. The combined application of electronics and optics has become common. An extreme example of electro-optics appears in some space cameras, in which the film is exposed, processed, and then scanned by a tiny point of light; the light passing through the film is picked up by a photocell and transmitted to Earth by radio, where it is made to control the brightness of another point of light scanning a second piece of film in exact synchronism with the scanning spot in the camera. The whole system thus produces a picture on Earth that is an exact replica of the picture photographed in space a few minutes earlier.

Projection Screens

The simplest screen for the projection of slides or motion pictures is, of course, a matte white surface, which may be on a hard base as in outdoor theatres or on a stretched cloth indoors. A theatre screen is often perforated to transmit sound from loudspeakers placed behind it.

Improved screen materials have been developed to increase the brightness of the picture to suit the particular shape of the auditorium. A screen covered with tiny beads tends to send the light back in the general direction of the projector, and is suitable for use at one end of a long, narrow auditorium. Another type of screen is covered with fine embossed vertical grooves; this tends to distribute the light in a horizontal band across the audience with little or no vertical spread. A real advantage of these highly reflective screens is that they tend to reflect ambient room light away from the viewer as by a mirror, so that the pictures appear almost as bright and clear by day as in a darkened room.

Reflecting Prisms

Reflecting prisms are pieces of glass bounded by plane surfaces set at carefully specified angles. Some of these surfaces transmit light some reflect light, while some serve both functions in succession. A prism is thus an assembly of plane reflectors at relatively fixed angles, which are traversed in succession by a beam of light. The simplest prism is a triangular block of glass with two faces at right angles and one at an angle of 45°. The face at 45° deflects a beam of light through a right angle. The common Porro prism used in a pair of binoculars contains four 45° reflecting surfaces, two to reverse the beam direction in the vertical plane and two in the horizontal plane. These reflecting faces could be replaced by pieces of mirror mounted on a metal frame, but it is hard to hold mirrors rigidly and harder still to keep them clean. Some microscopes are equipped with a 45° deflection prism behind the eyepiece; this prism may provide two or three reflections depending on the type of image inversion or left-for-right reversal required.

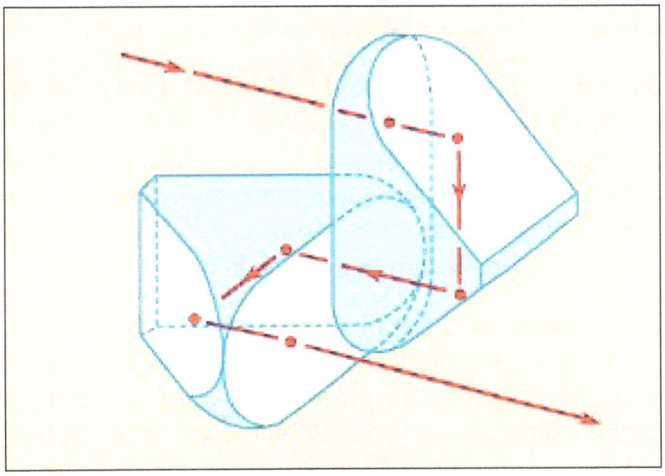

Porro prism.

Prisms containing a semi-reflecting, semi-transmitting surface are known as beam splitters and as such have many uses. An important application is found in some colour television cameras, in which the light from the lens is divided by two beam splitters in succession to form red, green, and blue images on the faces of three image tubes in the camera.

Dispersing Devices

There are two forms of dispersing element used to spread out the constituent colours of a beam of light into a "spectrum," namely a prism and a grating. The prism, known to Newton, is the older; it separates the colours of the spectrum because the refractive index of the glass is lowest for red light and progressively increases through the yellow and green to the blue, where it is highest. Prism spectroscopes and spectrographs are made in a variety of forms and sizes, but in all cases the blue end of the spectrum is greatly spread out while the red end is relatively compressed.

A diffraction grating is a ruled mirror or transparent plate of glass having many thousands of fine parallel grooves to the inch. It separates the colours of the spectrum by a process of diffraction. Each groove diffracts, or scatters, light in all directions, and in the case of light of one particular wavelength, there will be one direction in which the light wave from one groove lags behind the light wave from the next groove by precisely one or more whole wavelengths. This results in a strong beam of diffracted light in that direction and darkness in all other directions. Since each spectral colour corresponds to a different wavelength, the grating spreads out the spectrum into a fan where it can be observed or photographed. The red rays are bent most and the blue rays least, the opposite of the situation with a prism.

Although a prism or grating is the essential dispersing element in a spectrograph, a fine slit and additional lenses or focusing mirrors must be used to form a sharply defined spectrum. Prism spectroscopes are, of course, limited to those wavelengths for which the prism material is transparent; a reflecting grating can be used for any wavelength that the material will reflect.

Filters and Thin Films

A colour filter is a sheet of transparent material that modifies a light beam by selective absorption of some colours in relation to others. A neutral filter absorbs all wavelengths equally and merely serves to reduce the intensity of a beam of light without changing its colour.

Filters may be made from sheets of coloured glass, plastic, or dyed gelatin, and in some cases glass cells filled with liquid have been used. Since World War II, another type of filter depending on the interference of light has been developed in which one or more metallic or other types of films of controlled thickness have been deposited on a glass plate, the layers being so thin as to cause selective interference of some wavelengths in relation to others and thus act as a non-absorbing filter. In this case the rejected colours are reflected instead of being absorbed.

Polarizing filters have the property of transmitting light that vibrates in one direction while absorbing light that vibrates in a perpendicular direction. These filters are used extensively in scientific instruments. In sunglasses and when placed over a camera lens, polarizing filters reduce unwanted reflections from nonmetallic surfaces. Polarizing spectacles have been used to separate the left-eye and right-eye beams in the projection of stereoscopic pictures or movies.

Fibre-optics Bundles

A thin rod or fibre of glass or other transparent material transmits light by repeated internal reflections, even when the rod is somewhat curved. An ordered bundle of rods or fibres is thus capable of taking an image projected upon one end of the bundle and reproducing it at the other end. A fibre-optics bundle can be fused together into a rigid channel, or it may be left flexible, only the ends being rigidly fastened together. Because a fibre bundle is exceedingly delicate, it must be handled with care; breaking a fibre would cause a black dot to appear in the reproduced image.

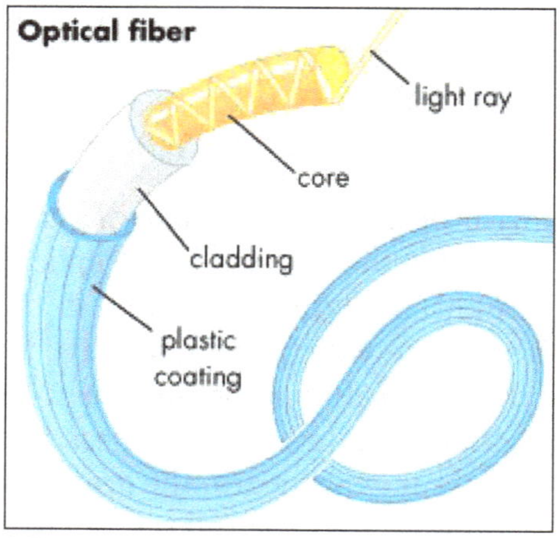

Optical fibre: Light ray passing through an optical fibre.

Nonclassical Imaging Systems

Besides the familiar optical systems cited above, there are many nonclassical optical elements that are used to a limited extent for special purposes. The most familiar of these is the aspheric (non-spherical) surface. Because plane and spherical surfaces are the easiest to generate accurately on glass, most lenses contain only such surfaces. It is occasionally necessary, however, to use some other axially symmetric surface on a lens or mirror, generally to correct a particular aberration. An example is the parabolic surface used for the primary mirror of a large astronomical telescope; another is the elliptic surface molded on the front of the little solid glass reflector units used on highway signs.

Another commonly used optical surface is the side of a cylinder. Such surfaces have power only in the meridian perpendicular to the cylinder axis. Cylindrical lenses are therefore used wherever it is desired to vary the magnification from one meridian to a perpendicular meridian. Cylindrical surfaces are employed in the anamorphic lenses used in some wide-screen motion-picture systems to compress the image horizontally in the camera and stretch it back to its original shape in the projected image.

To correct astigmatism in the eye, many spectacles are made with toric surfaces—i.e., with a stronger curvature in one meridian than in the perpendicular meridian, like the bowl of a teaspoon. These surfaces are generated and polished by special machines and are made by the million every year. Another nonclassical optical system is the bifocal or trifocal spectacle lens. They are made either by forming two or three separate surfaces on a single piece of glass or obtaining additional power by fusing a piece of high-index glass on to the front of the main lens and then polishing a single spherical surface over both glasses.

Two French scientists, Georges-Louis Buffon and Augustin-Jean Fresnel, in the 18th century suggested forming a lens in concentric rings to save weight, each ring being a portion of what would normally be a continuous spherical surface but flattened out. On a large scale, Fresnel lenses have been used in lighthouses, floodlights, and traffic signals, and as cylindrical ship's lanterns. With fine steps a few thousandths of an inch wide, molded plastic Fresnel lenses are often used as condensers in overhead projectors and in cameras as a field lens in contact with a ground-glass viewing screen.

Lenses have occasionally been made with one surface taking the form of a flattened cone. Such lenses produce a long, linear image of a point source, lying along the lens axis; for this reason they are commonly referred to as axicons. They have been used to produce a straight line of light in space for aligning machines and shafting, but since about 1965 the beam from a gas laser has generally been used instead.

Lens Aberrations

Seidel Sums

If a lens were perfect and the object were a single point of monochromatic light, then, as noted above, the light wave emerging from the lens would be a portion of a sphere centered about the ideal image point, lying in the paraxial image plane at a height above the axis given by the Lagrange theorem. In practice, however, this condition is most unlikely to occur; it is much more probable that the emerging wave will depart slightly from a perfect sphere, the departure varying from point to point over the lens aperture. This departure is extremely small, being of the order of the wavelength of light that is only half a micron, so it would be impossible to show this departure on a drawing. It can be represented mathematically, however, in the following way: The coordinates of a point in the exit-pupil aperture will be represented by x_o and y_o, the y_o coordinate lying in the meridian plane containing the object point and the lens axis. The departure of the wave from the ideal sphere is generally called OPD, meaning optical path difference. It can be shown that OPD is related to x_o and y_o by five constants S_1 through S_5, and the quantity h'_o,

$$\mathrm{OPD} = S_1\left(x_o^2+y_o^2\right)+S_2y_o\left(x_o^2+y_o^2\right)h'_o+S_3\left(x_o^2+3y_o^2\right)h'^2_o+S_4\left(x_o^2+y_o^2\right)h'^2_o+S_5y_oh'^3_o.$$

Each of these five terms is considered to be a separate "aberration," the coefficients S_1,...S_5, being called Seidel sums after the 19th-century German scientist L.P. Seidel, who identified the imperfections. These aberrations are respectively spherical, coma, astigmatism, Petzval field curvature, and distortion. The symbol h'_o refers to the height of the final image point above the lens axis, and hence it defines the obliquity of the beam.

The five Seidel sums can be calculated by tracing a paraxial ray from object to image through the lens and by tracing also a paraxial principal ray from the centre of the aperture stop outward in both directions toward the object and image, respectively. The angle of incidence i and the ray slope angle u of each of these paraxial rays at each surface are then listed and inserted into the following expressions for the five sums. The angle u'_o represents the final emerging slope of the paraxial ray.

The calculation starts by determining the radius A of the exit pupil by $A=\sqrt{x_o{}^2+y_o{}^2}$ and also the quantity K at each surface by,

$$K=\frac{1}{2}yn\left(\frac{n}{n'}-1\right)(i-u')$$

The corresponding K_{pr} for the paraxial principal ray is also determined at each surface. Then, the five aberrations may be written,

$$S_1=\frac{1}{4A_4}\sum Ki^2 \quad S_2=\frac{1}{A^3h'_o}\sum Kii_{pr}$$

$$S_3=\frac{1}{2A^2h'^2_o}\sum Ki^2_{pr} \quad S_4=\frac{u'^2_o}{4A^2}\sum\frac{n-n'}{nn'r}$$

$$S_5=\frac{1}{Ah'^3_o}\sum\left[K_{pr}ii_{pr}+\frac{1}{2}h'_ou'_o\left(u'^2_{pr}-u^2_{pr}\right)\right]$$

To interpret these aberrations, the simplest procedure is to find the components x′, y′ of the displacement of a ray from the Lagrangian image point in the paraxial focal plane, by differentiating the OPD expression given above. The partial derivatives $\partial\mathrm{OPD}/\partial x_o$ and $\partial\mathrm{OPD}/\partial y_o$ represent respectively the components of the slope of the wave relative to the reference sphere at any particular point (x_0, y_0). Hence, because a ray is always perpendicular to the wave, the ray displacements in the focal plane can be found by,

$$x' = f\frac{\partial OPD}{\partial x_0} \text{ and } y' = f\frac{\partial OPD}{\partial y_0}$$

in which f is the focal length of the lens. The aggregation of rays striking the focal plane will indicate the kind of image that is characteristic of each aberration.

This procedure will be applied to each of the five aberration terms separately, assuming that all the other aberrations are absent. Obviously, in a perfect lens x′ and y′ are zero because OPD is zero. It must be remembered, however, that by using rays instead of waves, all fine-structure effects caused by diffraction will be lost, and only the macroscopic image structure will be retained.

Spherical Aberration

The first term in the OPD expression is $OPD = S_1\left(x_0^2 + y_0^2\right)^2$. Hence,

$$x' = f\frac{\partial OPD}{\partial x_0} = 4fA^2S_1 \cdot x_0 \text{ and } y' = f\frac{\partial OPD}{\partial y_0} = 4fA^2S_1 \cdot y_0.$$

These displacements can both be eliminated simultaneously by applying a longitudinal shift L to the focal plane. This changes x′ by $-Lx_0/f$ and y′ by $-Ly_0/f$; hence, if L is made equal to $4f^2A^2S_1$, both ray displacements vanish. The aberration, therefore, represents a condition in which each zone of the lens has a different focus along the axis, the shift of focus from the paraxial image being proportional to A^2. This is known as spherical aberration.

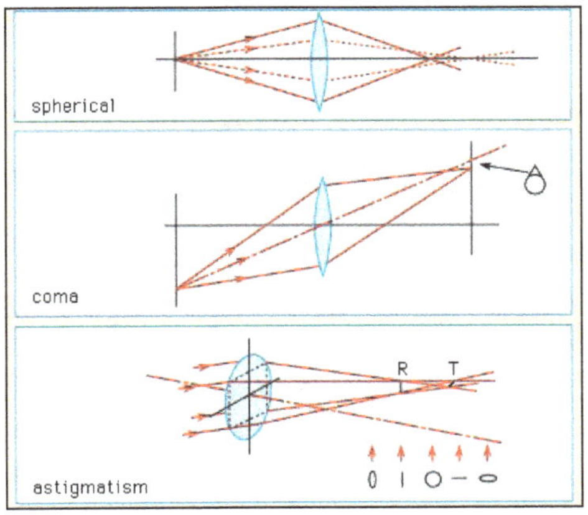

Lens aberrations.

Coma

The S_2 term in the OPD expression represents the aberration called coma, in which the image of a point has the appearance of a comet. The x′ and y′ components are as follows,

$$x' = fh'_o S_2 (2x_o y_o)$$
$$y' = fh'_o S_2 (x_o^2 + 3y_o^2)$$

When this aberration is present, each circular zone of the lens forms a small ringlike image in the focal plane, the rings formed by successive concentric zones of the lens fitting into two straight envelope lines at 60° to each other. Because the brightness of this image is greatest at the tip, coma tends to form a one-sided haze on images in the outer parts of the field.

Astigmatism

If only the S_3 term is present, then,

$$x' = 2fh_o'^2 S_3 (x_o)$$
$$y' = 2fh_o'^2 S_3 (3y_o)$$

For any one zone of the lens, x′ and y′ describe a vertical ellipse with major axis three times the minor axis. The images formed by all the smaller zones of the lens fit into this ellipse and fill it out with a uniform intensity of light. If the image plane is moved along the axis by a distance L, as in focussing a camera, then, $L = 2f^2 h_o'^2 S_3$, the ellipse shrinks to a radial focal line (R). Twice this displacement yields a circle; three times this L gives a tangential focal line (T), which is followed by an ellipse with its major axis in the x direction, bottom. The usual effect of astigmatism in an image is the appearance of radial or tangential blurring in the outer parts of the field.

Petzval Curvature

For the S_4 term taken alone,

$$x' = 2fh_o'^2 S_4 \cdot x_o$$
$$y' = 2fh_o'^2 S_4 \cdot y_o$$

The image of a point is now a small circle that contracts to a point at a new focus situated at a longitudinal distance $L = 2f^2 h_o'^2 S_4$, from the paraxial image. As the longitudinal displacement of the focus is proportional to the square of the image height h_o', this aberration represents a pure field curvature without any accompanying loss of definition (all lines remain sharp). It is named after the Hungarian mathematician József Petzval, who studied its properties in the early 1840s. The effect of Petzval curvature can be somewhat offset by the deliberate introduction of sufficient overcorrected astigmatism, as was done in all the pre-anastigmat photographic objectives. This added astigmatism is, of course, undesirable, and in order to design an anastigmat lens having a flat field free from astigmatism, it is necessary to reduce the Petzval sum S_4 drastically.

For a succession of thin lenses (1, 2, 3,...etc.) in a system, the Petzval sum becomes simply $1/f_1 n_1 + 1/f_2 n_2 + 1/f_3 n_3 +$...etc., in which f is the focal length of each element and n is its refractive index. Therefore, to reduce the sum and minimize this aberration, relatively strong negative elements of low-index glass can be combined with positive elements of high-index glass. The positive and negative elements must be axially separated to provide the lens with a useful amount of positive power. The introduction of high-index barium crown glass with a low dispersive power in the 1880s initiated the development of anastigmat lenses.

Distortion

For the S_5 aberration,

$$x' = 0$$

$$y' = fh_0'^3 S_5$$

When this aberration is present, the entire image point is displaced toward or away from the axis by an amount proportional to the third power of the transverse distance h_0' of the image from the axis. This leads to the formation of an image of a square that is either a barrel-shaped or a cushion-shaped figure.

It is to be noted that the five Seidel aberrations represent the largest and most conspicuous defects that can arise in an uncorrected optical system. Even in the best lenses in which these five aberrations have been perfectly corrected for one zone of the lens and for one point in the field, however, there will exist small residuals of these aberrations and of many other higher order aberrations also, which are significantly different from the classical types just described. The typical aberration figures shown in above figure are, of course, grossly exaggerated, and actually it requires some magnification of a star image to render these appearances clearly visible. Nevertheless, they are important enough to require drastic reduction in high-quality lenses intended to make sharp negatives capable of considerable enlargement.

Image Brightness

All photometric concepts are based on the idea of a standard candle, lamps having accurately known candle power being obtainable from the various national standards laboratories. The ratio of the candle power of a source to its area is called the luminance of the source; luminances range from about 2,000 candles per square millimetre at the surface of the Sun down to about 3×10^{-6} candle per square centimetre (3×10^{-6} stilb) for the luminous paint on a watch dial. Ordinary outdoor scenes in daylight have an average luminance of several hundred candles per square foot. The quantity of light flux flowing out from a source is measured in lumens, the lumen being defined as the amount of flux radiated by a small "point" source of one candle power into a cone having a solid angle of one steradian. When light falls upon a surface it produces illumination (i.e., illuminance), the usual measure of illuminance being the foot-candle, which is one lumen falling on each square foot of receiving surface.

It is often important to be able to calculate the brightness of an image formed by an optical system, because photographic emulsions and other light receptors cannot respond satisfactorily if the light level is too low. The problem is to relate the luminance of an object with the illuminance in the image, knowing the transmittance and aperture of the optical system. A small area A of a plane object having a luminance of B candles per square unit will have a normal intensity of AB candles. This source radiates light into a cone of semi-angle U, limited, for example, by the rim of a lens. The light flux (F) entering the cone can be found by integration to be,

$$F = \pi A B \sin^2 U \,\text{lumens}$$

If the object luminance is expressed as B_L lamberts, the lambert being an alternative luminance unit equal to 1/π (i.e., 0.32) candle per unit area, the flux (F) is,

$$F = A B_L \sin^2 U \,\text{lumens}$$

because there are π times as many lamberts in a given luminance as there are candles per unit area.

A fraction t of this flux finds its way to the image, t being the lens transmittance, generally about 0.8 or 0.9 but less if a mirror is involved. The area of the image is Am^2, in which m, the magnification, is given by,

$$m = \frac{\sin U}{\sin U'}$$

Hence, the image illuminance (E) is,

$$E = \left(t\pi AB\sin^2 U\right) \div \left(A\sin^2 U / \sin^2 U'\right) = t\pi B\sin^2 U' \text{ or}$$

$$E = tB_L \sin^2 U'$$

The image illuminance thus depends only on the luminance of the source and the cone angle of the beam proceeding from the lens to the image. This is a basic and most important relation underlying all calculations of image illuminance.

It is often more convenient to convert the angle U′ into other better known quantities, such as the f-number of the lens and the image magnification. The relation here is,

$$\sin U' = \frac{1}{2(f-\text{number})(1+m/m_p)}$$

The f-number of the lens is defined as the ratio of the focal length to the diameter of the entrance pupil; m is the image magnification; and m_p is the pupil magnification—i.e., the diameter of the exit pupil divided by the diameter of the entrance pupil. Combining above two equations gives,

$$\text{Image illuminance} = E = \frac{\pi t B}{4(f-\text{number})^2\left(1+m/m_p\right)^2}.$$

As an example in the use of this relation, if it is supposed that an f/2 lens is being used to project an image of a cathode-ray tube at five times magnification, the tube luminance being 5,000 foot-lamberts (1.7 candles per square centimetre), and the lens transmittance is 0.8, and the pupil magnification is unity. Then the image illuminance will be,

$$E = \frac{0.8 \times 5{,}000}{4 \times 4 \times 36} = 6.9 \,\text{foot-candles}$$

The image is very much less bright than the object, a fact that becomes clear to anyone attempting to provide a bright projected image in a large auditorium.

Distribution of Illumination Over an Image

So far only the illumination at the center of an image has been considered, but the distribution of illumination over a wide field is often important. In the absence of any lens, the small plane source already considered radiates in a direction inclined at an angle φ to the axis with an intensity AB cos φ. This light has to travel farther than the axial light to reach a screen, and then it strikes the screen at another angle φ. The net result is that the oblique illumination on the screen is smaller than the axial illumination by the factor $\cos^4 \varphi$.

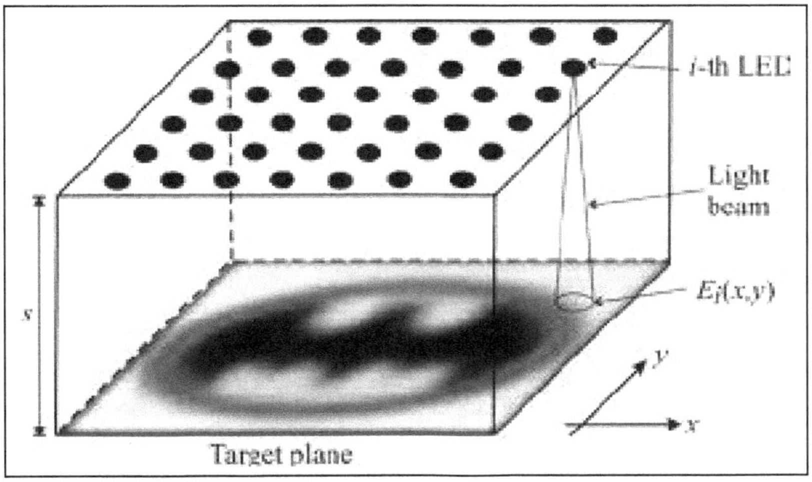

The same law can be applied to determine the oblique illumination due to a lens, assuming a uniform extended diffusing source of light on the other side of the lens. In this case, however, the exit pupil will not in general be a perfect circle because of possible distortion of the iris by that part of the optical system lying between the iris and the image. Also, any mechanical vignetting in the lens will make the aperture noncircular and reduce still further the oblique illumination. In a camera this reduction in oblique illumination results in darkened corners of the picture, but, if the reduction in brightness is gradual, it is not likely to be detected because the eye adapts quickly to changing brightness as the eyes scan over the picture area. Indeed, a 50 percent drop in brightness between the centre and corners of an ordinary picture is scarcely detectable.

Visual Brightness

The apparent brightness of things seen by the eye follows the same laws as any other imaging system, because the apparent brightness is measured by the illuminance in the image that is projected on the retina. The angle U′ in equation $E = \left(t\pi AB\sin^2 U\right) \div \left(A\sin^2 U / \sin^2 U'\right) = t\pi B\sin^2 U'$ or $E = tB_L \sin^2 U'$ inside the eye is determined by the size of the pupil of the eye, which varies from about one millimetre to about eight millimetres, depending on the brightness of the environment. Apart from this variation, retinal illuminance is directly proportional to object luminance, and objects having the same luminance appear equally bright, no matter at what distance they are observed.

From this argument, it is clear that no visual instrument, such as a telescope, can possibly make anything appear brighter than when viewed directly. To be sure, a telescope having a large objective lens accepts more light from an object in proportion to the area of the lens aperture, but it magnifies the image area in the same proportion; so the increased light is spread over an increased area of the retina, and the illuminance remains unchanged. Actually, the telescopic view is always dimmer than the direct view because of light losses in the telescope due to glass absorption and surface reflections and because the exit pupil of the telescope may be smaller than the pupil of the eye, thus reducing the angle U′.

The case of a star being observed through a telescope is quite different, because no degree of magnification can possibly make a star appear as anything other than a point of light. Hence, star images appear brighter in proportion to the area of the telescope objective (assuming that the exit pupil is larger than the eye pupil), and the visibility of a star against the sky background is thus improved in proportion to the square of the diameter of the telescope objective lens.

Atmospheric Optics

A colorful sky is often due to scattering of light off particulates, as in this photograph of a sunset during the October 2007 California wildfires.

Atmospheric optics deals with how the unique optical properties of the Earth's atmosphere cause a wide range of spectacular optical phenomena. The blue color of the sky is a direct result of Rayleigh scattering which redirects higher frequency (blue) sunlight back into the field of view of the observer. Because blue light is scattered more easily than red light, the sun takes on a reddish hue when it is observed through a thick atmosphere, as during a sunrise or sunset. Additional particulate matter in the sky can scatter different colors at different angles creating colorful glowing skies at dusk and dawn. Scattering off of ice crystals and other particles in the atmosphere are responsible for halos, afterglows, coronas, rays of sunlight, and sun dogs. The variation in these kinds of phenomena is due to different particle sizes and geometries.

Mirages are optical phenomena in which light rays are bent due to thermal variations in the refraction index of air, producing displaced or heavily distorted images of distant objects. Other optical phenomena associated with this include the Novaya Zemlya effect where the sun appears to rise earlier or set later than predicted with a distorted shape. A spectacular form of refraction occurs with a temperature inversion called the Fata Morgana where objects on the horizon or even beyond the horizon, such as islands, cliffs, ships or icebergs, appear elongated and elevated, like "fairy tale castles".

Rainbows are the result of a combination of internal reflection and dispersive refraction of light in raindrops. Because rainbows are seen on the opposite side of the sky as the sun, rainbows are more prominent the closer the sun is to the horizon due to their greater distance apart.

Sun and Moon Size

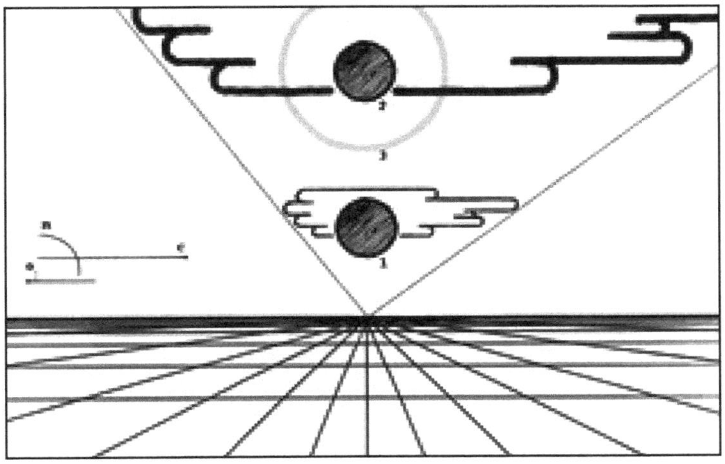

Comparison between the relative sizes of the Moon and a cloud at different points in the sky.

In the Book of Optics (1011–1022 A.D.), Ibn al-Haytham argued that vision occurs in the brain, and that personal experience has an effect on what people see and how they see, and that vision and perception are subjective. Arguing against Ptolemy's refraction theory for why people perceive the sun and moon larger at the horizon than when they are higher in the sky, he redefined the problem in terms of perceived, rather than real, enlargement. He said that judging the distance of an object depends on there being an uninterrupted sequence of intervening bodies between the object and the observer. With the Moon, however, there are no intervening objects. Therefore, since the size of an object depends on its observed distance, which is in this case inaccurate, the Moon appears larger on the horizon. Through works by Roger Bacon, John Pecham and Witelo based on Ibn al-Haytham's explanation, the Moon illusion gradually came to be accepted as a psychological phenomenon, with Ptolemy's theory being rejected in the 17th century. For over 100 years, research on the Moon illusion has been conducted by vision scientists who invariably have been psychologists specializing in human perception. After reviewing the many different explanations in their 2002 book *The Mystery of the Moon Illusion*, Ross and Plug conclude "No single theory has emerged victorious".

Sky Coloration

Light from the sky is a result of the scattering of sunlight, which results in a blue color perceived by the human eye. On a sunny day Rayleigh scattering gives the sky a blue gradient — dark in the zenith, light near the horizon. Light that comes in from overhead encounters 1/38th of the air mass that light coming along a horizon path encounters. So, fewer particles scatter the zenith sunbeam, and therefore the light remains a darker blue. The blueness is at the horizon because the blue light coming from great distances is also preferentially scattered. This results in a red shift of the far lightsources that is compensated by the blue hue of the scattered light in the line of sight. In other words, the red light scatters also; if it does so at a point a great distance from the observer it has a much higher chance of reaching the observer than blue light. At distances nearing infinity the scattered light is therefore white. Far away clouds or snowy mountaintops will seem yellow for that reason; that effect is not obvious on clear days, but very pronounced when clouds are covering the line of sight reducing the blue hue from scattered sunlight.

When seen from altitude, as here from an airplane, the sky's color varies from pale to dark at elevations approaching the zenith.

The scattering due to molecule sized particles (as in air) is greater in the forward and backward directions than it is in the lateral direction. Individual water droplets exposed to white light will create a set of colored rings. If a cloud is thick enough, scattering from multiple water droplets will wash out the set of colored rings and create a washed out white color. Dust from the Sahara moves around the southern periphery of the subtropical ridge moves into the southeastern United States during the summer, which changes the sky from a blue to a white appearance and leads to an increase in red sunsets. Its presence negatively impacts air quality during the summer since it adds to the count of airborne particulates.

Purple sky on the La Silla Observatory.

The sky can turn a multitude of colors such as red, orange, pink and yellow (especially near sunset or sunrise) and black at night. Scattering effects also partially polarize light from the sky, most pronounced at an angle 90° from the sun.

Sky luminance distribution models have been recommended by the International Commission on Illumination (CIE) for the design of daylighting schemes. Recent developments relate to "all sky models" for modelling sky luminance under weather conditions ranging from clear sky to overcast.

Cloud Coloration

An occurrence of altocumulus and cirrocumulus cloud iridescence.

The color of a cloud, as seen from the Earth, tells much about what is going on inside the cloud. Dense deep tropospheric clouds exhibit a high reflectance (70% to 95%) throughout the visible spectrum. Tiny particles of water are densely packed and sunlight cannot penetrate far into the cloud before it is reflected out, giving a cloud its characteristic white color, especially when viewed from the top. Cloud droplets tend to scatter light efficiently, so that the intensity of the solar radiation decreases with depth into the gases. As a result, the cloud base can vary from a very light to very dark grey depending on the cloud's thickness and how much light is being reflected or transmitted back to the observer. Thin clouds may look white or appear to have acquired the color of their environment or background. High tropospheric and non-tropospheric clouds appear mostly white if composed entirely of ice crystals and/or supercooled water droplets.

As a tropospheric cloud matures, the dense water droplets may combine to produce larger droplets, which may combine to form droplets large enough to fall as rain. By this process of accumulation, the space between droplets becomes increasingly larger, permitting light to penetrate farther into the cloud. If the cloud is sufficiently large and the droplets within are spaced far enough apart, it may be that a percentage of the light which enters the cloud is not reflected back out before it is absorbed. A simple example of this is being able to see farther in heavy rain than in heavy fog. This process of reflection/absorption is what causes the range of cloud color from white to black.

Other colors occur naturally in clouds. Bluish-grey is the result of light scattering within the cloud. In the visible spectrum, blue and green are at the short end of light's visible wavelengths, while red and yellow are at the long end. The short rays are more easily scattered by water droplets, and the long rays are more likely to be absorbed. The bluish color is evidence that such scattering is being produced by rain-sized droplets in the cloud. A greenish tinge to a cloud is produced when sunlight is scattered by ice. A cumulonimbus cloud emitting green is a sign that it is a severe thunderstorm, capable of heavy rain, hail, strong winds and possible tornadoes. Yellowish clouds may occur in the late spring through early fall months during forest fire season. The yellow color is due to the presence of pollutants in the smoke. Yellowish clouds caused by the presence of nitrogen dioxide are sometimes seen in urban areas with high air pollution levels.

Sunset reflecting shades of pink onto grey stratocumulus clouds.

Red, orange and pink clouds occur almost entirely at sunrise and sunset and are the result of the scattering of sunlight by the atmosphere. When the angle between the sun and the horizon is less than 10 percent, as it is just after sunrise or just prior to sunset, sunlight becomes too red due to refraction for any colors other than those with a reddish hue to be seen. The clouds do not become that color; they are reflecting long and unscattered rays of sunlight, which are predominant at those hours. The effect is much like if one were to shine a red spotlight on a white sheet. In combination with large, mature thunderheads this can produce blood-red clouds. Clouds look darker in the near-infrared because water absorbs solar radiation at those wavelengths.

Halos

A halo (λως; also known as a nimbus, icebow or gloriole) is an optical phenomenon produced by the interaction of light from the sun or moon with ice crystals in the atmosphere, resulting in colored or white arcs, rings or spots in the sky. Many halos are positioned near the sun or moon, but others are elsewhere and even in the opposite part of the sky. They can also form around artificial lights in very cold weather when ice crystals called diamond dust are floating in the nearby air.

There are many types of ice halos. They are produced by the ice crystals in cirrus or cirrostratus clouds high in the upper troposphere, at an altitude of 5 kilometres (3.1 mi) to 10 kilometres (6.2 mi), or, during very cold weather, by ice crystals called diamond dust drifting in the air at low levels. The particular shape and orientation of the crystals are responsible for the types of halo observed. Light is reflected and refracted by the ice crystals and may split into colors because of dispersion. The crystals behave like prisms and mirrors, refracting and reflecting sunlight between their faces, sending shafts of light in particular directions. For circular halos, the preferred angular distance are 22 and 46 degrees from the ice crystals which create them. Atmospheric phenomena such as halos have been used as part of weather lore as an empirical means of weather forecasting, with their presence indicating an approach of a warm front and its associated rain.

A man in front of a complex halo display.

Sun Dogs

Very bright sundogs in Fargo, North Dakota. Note the halo arcs passing through each sun dog.

Sun dogs are a common type of halo, with the appearance of two subtly-colored bright spots to the left and right of the sun, at a distance of about 22° and at the same elevation above the horizon.

They are commonly caused by plate-shaped hexagonal ice crystals. These crystals tend to become horizontally aligned as they sink through the air, causing them to refract the sunlight to the left and right, resulting in the two sun dogs.

As the sun rises higher, the rays passing through the crystals are increasingly skewed from the horizontal plane. Their angle of deviation increases and the sundogs move further from the sun. However, they always stay at the same elevation as the sun. Sun dogs are red-colored at the side nearest the sun. Farther out the colors grade to blue or violet. However, the colors overlap considerably and so are muted, rarely pure or saturated. The colors of the sun dog finally merge into the white of the parhelic circle (if the latter is visible).

It is theoretically possible to predict the forms of sun dogs as would be seen on other planets and moons. Mars might have sundogs formed by both water-ice and CO_2-ice. On the giant gas planets — Jupiter, Saturn, Uranus and Neptune — other crystals form the clouds of ammonia, methane, and other substances that can produce halos with four or more sundogs.

Glory

Solar glory at the steam from a hot spring .

A common optical phenomenon involving water droplets is the glory. A glory is an optical phenomenon, appearing much like an iconic Saint's halo about the head of the observer, produced by light backscattered (a combination of diffraction, reflection and refraction) towards its source by a cloud of uniformly sized water droplets. A glory has multiple colored rings, with red colors on the outermost ring and blue/violet colors on the innermost ring.

The angular distance is much smaller than a rainbow, ranging between 5° and 20°, depending on the size of the droplets. The glory can only be seen when the observer is directly between the sun and cloud of refracting water droplets. Hence, it is commonly observed while airborne, with the glory surrounding the airplane's shadow on clouds (this is often called *The Glory of the Pilot*). Glories can also be seen from mountains and tall buildings, when there are clouds or fog below the level of the observer, or on days with ground fog. The glory is related to the optical phenomenon anthelion.

Rainbow

Double rainbow and supernumerary rainbows on the inside of the primary arc. The shadow of the photographer's head marks the centre of the rainbow circle (antisolar point).

A rainbow is an optical and meteorological phenomenon that causes a spectrum of light to appear in the sky when the Sun shines on to droplets of moisture in the Earth's atmosphere. It takes the form of a multicolored arc. Rainbows caused by sunlight always appear in the section of sky directly opposite the sun, but originate no further than 42 degrees above the horizon for observers on the ground. To see them at higher angles, an observer would need to be in an airplane or near a mountaintop since the rainbow would otherwise be below the horizon. The bigger the droplets which formed the rainbow, the brighter it will be. Rainbows are most common near afternoon thunderstorms during the summer.

A single reflection off the backs of an array of raindrops produces a rainbow with an angular size on the sky that ranges from 40° to 42° with red on the outside. Double rainbows are produced by two internal reflections with angular size of 50.5° to 54° with violet on the outside. Within the "primary rainbow" (the lowest, and also normally the brightest rainbow) the arc of a rainbow shows red on the outer (or upper) part of the arc, and violet on the inner section. This rainbow is caused by light being reflected once in droplets of water. In a double rainbow, a second arc may be seen above and outside the primary arc, and has the order of its colors reversed (red faces inward toward the other rainbow, in both rainbows). This second rainbow is caused by light reflecting twice inside water droplets. The region between a double rainbow is dark. The reason for this dark band is that, while light *below* the primary rainbow comes from droplet reflection, and light *above* the upper (secondary) rainbow also comes from droplet reflection, there is no mechanism for the region *between* a double rainbow to show any light reflected from water drops, at all.

A rainbow spans a continuous spectrum of colors; the distinct bands (including the number of bands) are an artifact of human color vision, and no banding of any type is seen in a black-and-white photograph of a rainbow (only a smooth gradation of intensity to a maxima, then fading to a minima at the other side of the arc). For colors seen by a normal human eye, the most commonly cited and remembered sequence, in English, is Newton's sevenfold red, orange, yellow, green, blue, indigo and violet (popularly memorized by mnemonics like Roy G. Biv). However, color-blind persons will see fewer colors.

Rainbows can be caused by many forms of airborne water. These include not only rain, but also mist, spray, and airborne dew.

Mirage

Various kinds of mirages in one location taken over the course of six minutes. The uppermost inset frame shows an inferior mirage of the Farallon Islands. The second inset frame shows a green flash on the left-hand side. The two lower frames and the main frame all show superior mirages of the Farallon Islands. In these three frames, the superior mirage evolves from a 3-image mirage to a 5-image mirage, and back to a 2-image mirage. Such a display is consistent with a Fata Morgana.

A mirage is a naturally occurring optical phenomenon in which light rays are bent to produce a displaced image of distant objects or the sky. The word comes to English via the French *mirage*, from the Latin *mirare*, meaning "to look at, to wonder at". This is the same root as for "mirror" and "to admire". Also, it has its roots in the Arabic *mirage*.

In contrast to a hallucination, a mirage is a real optical phenomenon which can be captured on camera, since light rays actually are refracted to form the false image at the observer's location. What the image appears to represent, however, is determined by the interpretive faculties of the human mind. For example, inferior images on land are very easily mistaken for the reflections from a small body of water.

Mirages can be categorized as "inferior" (meaning lower), "superior" (meaning higher) and "Fata Morgana", one kind of superior mirage consisting of a series of unusually elaborate, vertically stacked images, which form one rapidly changing mirage.

Green flashes and green rays are optical phenomena that occur shortly after sunset or before sunrise, when a green spot is visible, usually for no more than a second or two, above the sun, or a green ray shoots up from the sunset point. Green flashes are actually a group of phenomena stemming from different causes, and some are more common than others. Green flashes can be observed from any altitude (even from an aircraft). They are usually seen at an unobstructed horizon, such as over the ocean, but are possible over cloud tops and mountain tops as well.

A green flash from the moon and bright planets at the horizon, including Venus and Jupiter, can also be observed.

Fata Morgana

A Fata Morgana of a boat.

This optical phenomenon occurs because rays of light are strongly bent when they pass through air layers of different temperatures in a steep thermal inversion where an atmospheric duct has formed. A thermal inversion is an atmospheric condition where warmer air exists in a well-defined layer above a layer of significantly cooler air. This temperature inversion is the opposite of what is normally the case; air is usually warmer close to the surface, and cooler higher up. In calm weather, a layer of significantly warmer air can rest over colder dense air, forming an atmospheric duct which acts like a refracting lens, producing a series of both inverted and erect images.

A Fata Morgana is an unusual and very complex form of mirage, a form of superior mirage, which, like many other kinds of superior mirages, is seen in a narrow band right above the horizon. It is an Italian phrase derived from the vulgar Latin for "fairy" and the Arthurian sorcerer Morgan le Fay, from a belief that the mirage, often seen in the Strait of Messina, were fairy castles in the air, or false land designed to lure sailors to their death created by her witchcraft. Although the term Fata Morgana is sometimes incorrectly applied to other, more common kinds of mirages, the true Fata Morgana is not the same as an ordinary superior mirage, and is certainly not the same as an inferior mirage.

Fata Morgana mirages tremendously distort the object or objects which they are based on, such that the object often appears to be very unusual, and may even be transformed in such a way that it is completely unrecognizable. A Fata Morgana can be seen on land or at sea, in polar regions or in deserts. This kind of mirage can involve almost any kind of distant object, including such things as boats, islands, and coastline.

A Fata Morgana is not only complex, but also rapidly changing. The mirage comprises several inverted (upside down) and erect (right side up) images that are stacked on top of one another. Fata Morgana mirages also show alternating compressed and stretched zones.

Novaya Zemlya Effect

The Novaya Zemlya effect is a polar mirage caused by high refraction of sunlight between atmospheric thermoclines. The Novaya Zemlya effect will give the impression that the sun is rising earlier or setting later than it actually should (astronomically speaking). Depending on the meteorological situation the effect will present the sun as a line or a square (which is sometimes referred to as the "rectangular sun"), made up of flattened hourglass shapes. The mirage requires rays of sunlight to have an inversion layer for hundreds of kilometres, and depends on the inversion layer's temperature gradient. The sunlight must bend to the Earth's curvature at least 400 kilometres (250 mi) to allow an elevation rise of 5 degrees for sight of the sun disk.

The first person to record the phenomenon was Gerrit de Veer, a member of Willem Barentsz' ill-fated third expedition into the polar region. Novaya Zemlya, the archipelago where de Veer first observed the phenomenon, lends its name to the effect.

Crepuscular Rays

Crepuscular rays, taken near Waterberg Plateau, Namibia.

Crepuscular rays are near-parallel rays of sunlight moving through the Earth's atmosphere, but appear to diverge because of linear perspective. They often occur when objects such as mountain peaks or clouds partially shadow the sun's rays like a cloud cover. Various airborne compounds scatter the sunlight and make these rays visible, due to diffraction, reflection, and scattering.

Crepuscular rays can also occasionally be viewed underwater, particularly in arctic areas, appearing from ice shelves or cracks in the ice. Also they are also viewed in days when the sun hits the clouds in a perfect angle shining upon the area.

There are three primary forms of crepuscular rays:

- Rays of light penetrating holes in low clouds (also called "Jacob's Ladder").
- Beams of light diverging from behind a cloud.
- Pale, pinkish or reddish rays that radiate from below the horizon. These are often mistaken for light pillars.

They are commonly seen near sunrise and sunset, when tall clouds such as cumulonimbus and mountains can be most effective at creating these rays.

Anticrepuscular Rays

Anticrepuscular rays while parallel in reality are sometimes visible in the sky in the direction opposite the sun. They appear to converge again at the distant horizon.

Atmospheric Refraction

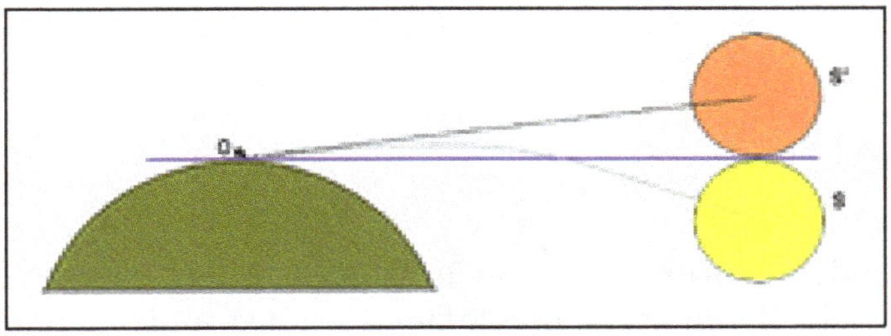

Diagram showing displacement of the Sun's image at sunrise and sunset.

Atmospheric refraction influences the apparent position of astronomical and terrestrial objects, usually causing them to appear higher than they actually are. For this reason navigators, astronomers, and surveyors observe positions when these effects are minimal. Sailors will only shoot a star when 20° or more above the horizon, astronomers try to schedule observations when an object is highest in the sky, and surveyors try to observe in the afternoon when refraction is minimum.

References

- Optics, physics: byjus.com, Retrieved 21, January 2020
- Basics-of-Fiber-Optics: fibersystems.com, Retrieved 10, August 2020

2

Physical Optics

Physical optics is the study of various phenomena of light where it behaves like electromagnetic wave and not like a particle. Some of the phenomena studied under physical optics are superposition, interference, diffraction, dispersion and polarization. This chapter discusses in detail the theories and phenomena related to physical optics.

Physical optics is the study of the physical properties and behavior of electromagnetic waves and their interactions with matter. The optical range of wavelengths may be taken as the range from about 1 nanometer (4×10^{-8} in.) to about 1 millimeter (0.04 in.). More narrowly, physical optics deals with the relationship between the atomic structure of a system and the absorption, emission, and propagation of light inside the system. The chief founder of this branch of science was Michael Faraday, who in 1845 provided the first clue to the electromagnetic nature of light by showing that the optical properties of glass could be altered by a magnetic field. Branches of physical optics include wave optics and Fourier optics. Fourier optics utilizes the mathematical technique of the Fourier transform to analyze the propagation of light and the formation of images.

Superposition

The principle of superposition says: When two or more waves cross at a point, the displacement at that point is equal to the sum of the displacements of the individual waves. The individual wave is placements may be positive or negative. If the displacements are vectors, then the sum is calculated by vector addition.

Superposition is an important idea that can explain phenomena including interference, diffraction and standing waves. It works for any type of wave (sound waves, water surface waves, electromagnetic waves) but only works under certain conditions, which we describe below. When the waves pass beyond a point of intersection, they separate out again and are unaffected.

Conditions for Superposition

The principle of superposition can be applied to any type of wave providing that: The waves being superposed are of the same type (e.g. all are electromagnetic waves). The medium that the waves are propagating through behaves linearly, i.e. when part of the medium has twice the displacement then it has twice the restoring force. This is usually true when the amplitudes are relatively small. For example, for waves on water, it is a good approximation for small ripples on a pond whose amplitude is much smaller than their wavelength. If the waves are also coherent - i.e. if they all have the same frequency and a constant phase difference - then the superposition resembles another wave with the same frequency.

Coherence

If waves from two sources spread out and cross throughout a region of space then superposition occurs at every point, and a pattern of interference is obtained. The superposition of waves from two sources can usually only result in an observable fixed (stationary) interference pattern if the sources are coherent. This means that the waves from the sources have both the same frequency and the phase difference between them is constant.

For electromagnetic waves, this can only be achieved by dividing waves from a single coherent source, such as a laser, or by diffracting light from a single bulb so that it illuminates two slits. This is why we do not usually observe the interference of light in daily life. Coherent sources of sound waves can be set up easily, by connecting two or more speakers to the same signal generator. Four waves are passing through a point in space. Their variations at this point with time are shown below:

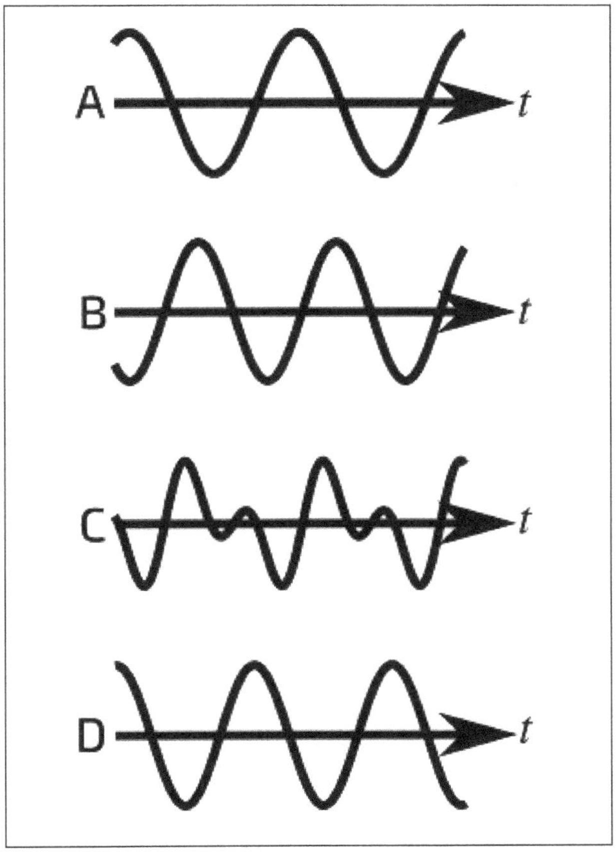

Four different waves, two of which are coherent.

Constructive and Destructive Interference

When two coherent waves are superposed, the amplitude of the resultant displacement (i.e. the peak value of displacement as it oscillates over time) depends on the phase difference between the two waves.

- If the two waves are in phase, there is constructive interference: The resultant amplitude is large, equal to the sum of the two wave amplitudes.
- If the two waves are in antiphase, there is destructive interference: The resultant amplitude is small, equal to the difference between the two wave amplitudes.
- In general, the two waves are neither in phase nor in antiphase: The resultant amplitude is somewhere in between the two extremes.

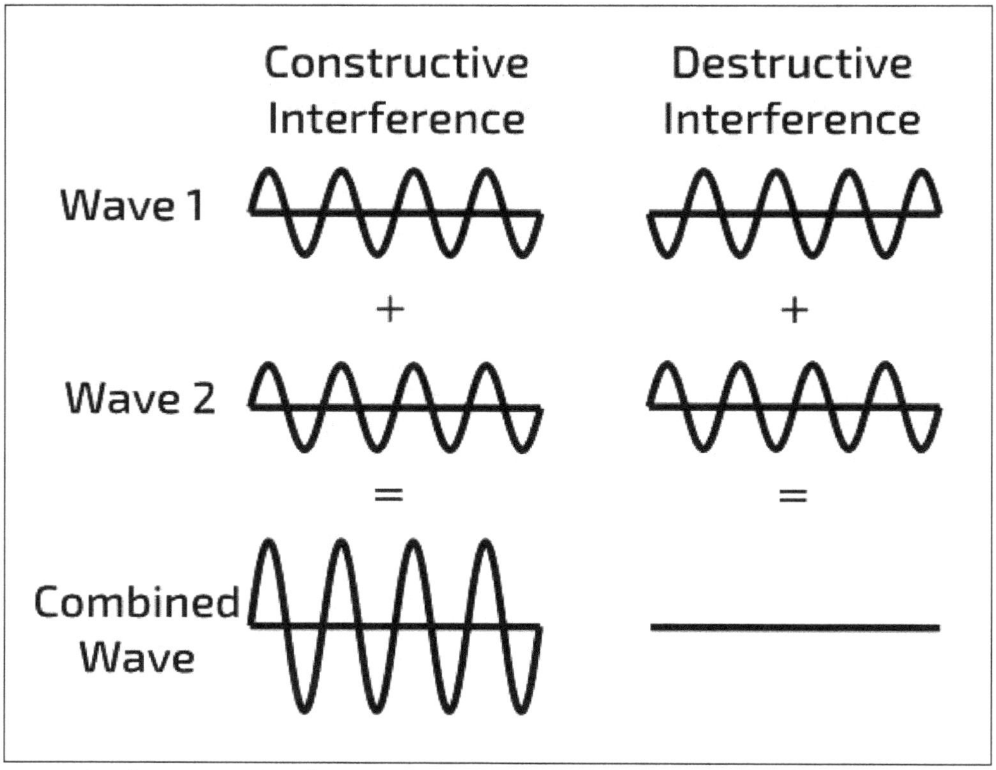

Graphs representing the constructive and destructive interference of two waves. Two waves in phase give constructive interference. Two waves pi, r, a, d, π rad out of phase give destructive interference.

Two coherent waves have a relative phase difference of pi, r, a, d, π rad (180, degrees,180°). The phase of one of them is increased by pi, r, a, d, π rad by reflection.

Path and Phase Difference

Phase is an angle, which gives the fraction of the progress through a repetitive cycle in terms of an angle between 0, 0 and 2, pi, 2 π radians or 0,0 and 360, degrees,360°. This progress through the cycle may be over a distance left bracket, lambda, right bracket,(λ) (the wavelength) or over time left bracket, T, right bracket, (T) (the period).

When two coherent waves superpose, the resultant wave depends on the phase difference between the two. If the waves, having been emitted with no phase difference between the two sources, have followed two different paths to the same point, then the phase difference is proportional to the difference in length between the paths, known as the path difference.

If the path difference is an integer multiple of the wavelength (lambda, comma, 2, lambda, comma, point, point, point, λ, 2 λ,...), then the waves are in phase and interfere constructively. If the path difference is a half-integer multiple of the wavelength (start fraction, lambda, divided by, 2, end fraction, comma, start fraction, 3, lambda, divided by, 2, end fraction, comma, point, point, point,2 λ,23 λ,...), then the waves are pi, r, a, d,πrad out of phase and interfere destructively if the sources are in phase. More generally, the link between a path difference of delta, x, Δx and a phase difference in radians of delta, $\Delta\Phi$ is:

$$\Delta\Phi = 2\pi\frac{\Delta x}{\lambda}$$

where λ is the wavelength of the wave in that medium. (To work in degrees, simply replace, 2 π by, 360).

For a wave moving through a material of refractive index n, the wavelength and the wavespeed are both reduced by a factor of n, so if the wavelength in a vacuum is lambda, start subscript, o, end subscript, λ_o:

$$\Delta\Phi = 2\pi n\frac{\Delta x}{\lambda_o}$$

What about waves that are not initially in phase? If the two waves have a phase difference α between them when they are emitted, then travel to the detector with a path difference of delta, Δx, the total phase difference between them at the detector is equal to:

$$\alpha + 2\pi\frac{\Delta x}{\lambda}$$

Two speakers at points A and B emit coherent sound of wavelength 1.4m and the sources have no phase difference between them. Point P is 10.8m from A and 8.7m from B.

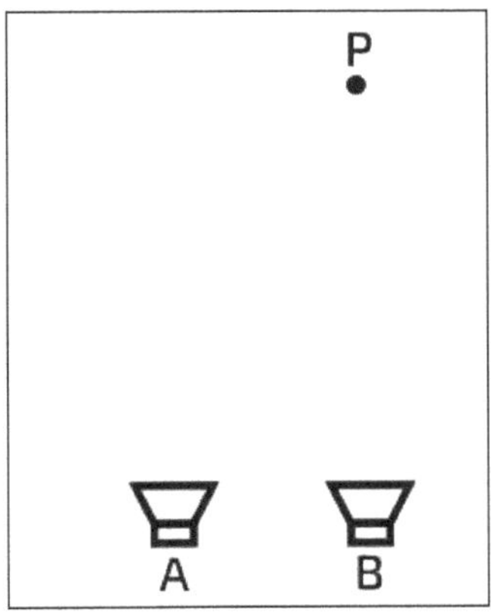

Speakers which give rise to interference.

Amplitude and Intensity

When superposition occurs, for slow mechanical waves the amplitude can usually be observed easily - for example, the height of ripples in a pond. For high frequency waves, such as sound or EM waves, the amplitude can be calculated by measuring the wave's intensity. Intensity is a measure of the energy of the wave in the region, which is proportional to the square of the amplitude:

$$\text{Intensity} \propto E \propto A^2$$

Beats

When two waves of slightly differing frequencies are superposed at some location, the resultant amplitude appears to vary periodically. This phenomenon is known as beating and is most commonly observed with sound waves.

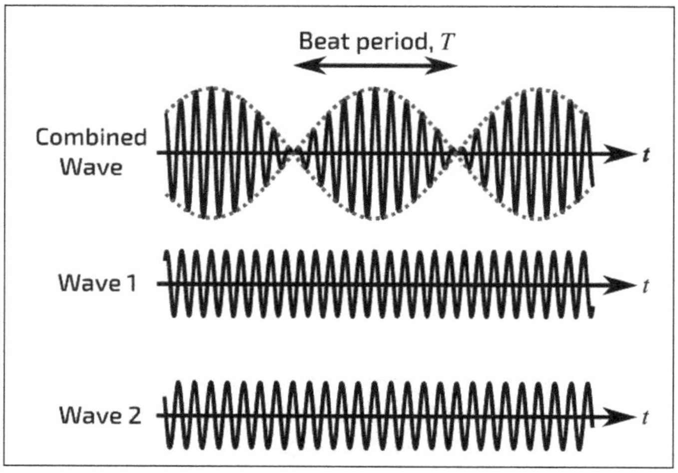

Two waves of similar frequency superpose, to give a wave showing beats. At the points of high amplitude in the combined wave, the two waves interfere constructively. At the points of low amplitude in the combined wave, the interference is destructive.

The beat frequency is $f_b = \frac{1}{T}$, where T, is the beat period as indicated above. If the two superposing waves have frequencies f_1, and f_2, where f_1, is the greater of the two, then the beat frequency is given by,

$$f_b = f_1 - f_2$$

Beat Frequency

The time variation of the two superposing waves of unit amplitude at frequencies f1, and f2 could be described mathematically by:

$$y1 = \cos(\omega_1 t) = \cos(2\pi f_1 t)$$
$$y2 = \cos(\omega_2 t) = \cos(2\pi f_2 t)$$

The cosine, $\left(2\pi\left(\frac{f_1+f_2}{2}\right)t\right)$ term gives a sinusoidal wave with a frequency which is the average of the two original frequencies. The cosine $\left(2\pi\left(\frac{f_1-f_2}{2}\right)t\right)$ term varies more slowly, and so gives the beat (amplitude) envelope.

Although the frequency of this envelope is $\frac{f_1 - f_2}{2}$, the beat period is the amount of time between consecutive minima in amplitude and the period of the cosine envelope is twice that. Whenever the amplitude of the $\frac{f_1 - f_2}{2}$ note heard becomes a maximum, the sound volume is loud, and this decides the beat frequency, which makes it twice the frequency of the envelope. So the beat frequency is $f_1 - f_2$.

Interference

Interference is defined as the phenomenon of superimposition of two or more waves of same frequency emitted by coherent sources such that the resultant wave is equal to the sum of the amplitude of the individual waves. Depending on whether the interference is constructive or destructive, the amplitude of the resultant wave can either be larger or smaller than the individual waves.

Wave interference is the phenomenon that occurs when two waves meet while traveling along the same medium. The interference of waves causes the medium to take on a shape that results from the net effect of the two individual waves upon the particles of the medium. To begin our exploration of wave interference, consider two pulses of the same amplitude traveling in different directions along the same medium. Let's suppose that each displaced upward 1 unit at its crest and has the shape of a sine wave. As the sine pulses move towards each other, there will eventually be a moment in time when they are completely overlapped. At that moment, the resulting shape of the medium would be an upward displaced sine pulse with an amplitude of 2 units. The diagrams below depict the before and during interference snapshots of the medium for two such pulses. The individual sine pulses are drawn in red and blue and the resulting displacement of the medium is drawn in green.

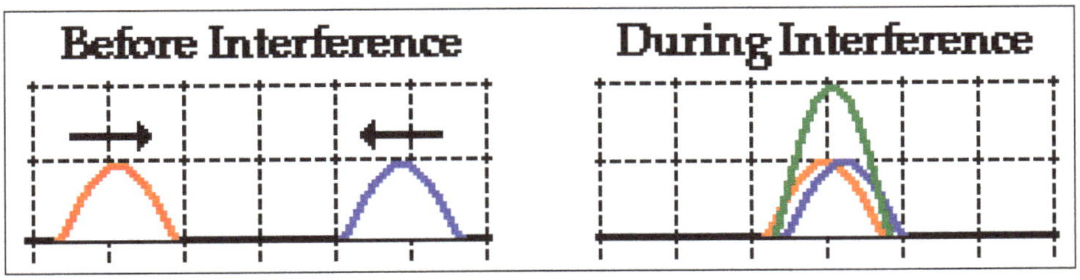

Constructive Interference

This type of interference is sometimes called constructive interference. Constructive interference is a type of interference that occurs at any location along the medium where the two interfering waves have a displacement in the same direction. In this case, both waves have an upward displacement; consequently, the medium has an upward displacement that is greater than the displacement of the two interfering pulses. Constructive interference is observed at any location where the two interfering waves are displaced upward. But it is also observed when both interfering waves are displaced downward. This is shown in the diagram below for two downward displaced pulses.

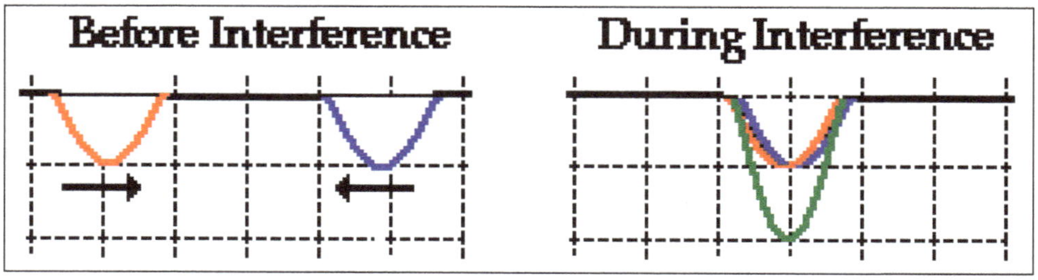

In this case, a sine pulse with a maximum displacement of -1 unit (negative means a downward displacement) interferes with a sine pulse with a maximum displacement of -1 unit. These two pulses are drawn in red and blue. The resulting shape of the medium is a sine pulse with a maximum displacement of -2 units.

Destructive Interference

Destructive interference is a type of interference that occurs at any location along the medium where the two interfering waves have a displacement in the opposite direction. For instance, when a sine pulse with a maximum displacement of +1 unit meets a sine pulse with a maximum displacement of -1 unit, destructive interference occurs. This is depicted in the diagram below.

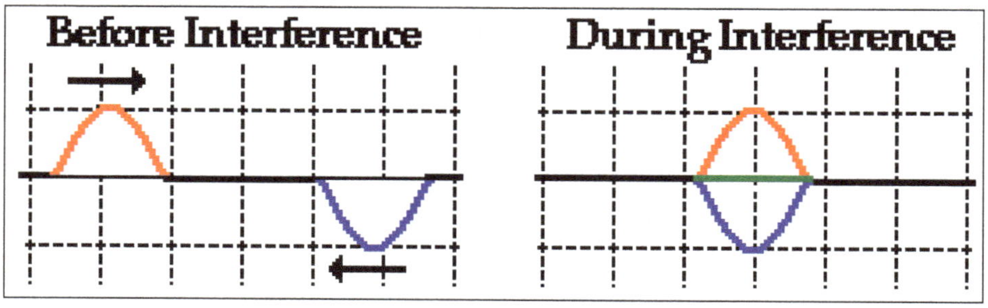

In the diagram above, the interfering pulses have the same maximum displacement but in opposite directions. The result is that the two pulses completely destroy each other when they are completely overlapped. At the instant of complete overlap, there is no resulting displacement of the particles of the medium. This "destruction" is not a permanent condition. In fact, to say that the two waves destroy each other can be partially misleading. When it is said that the two pulses destroy each other, what is meant is that when overlapped, the effect of one of the pulses on the displacement of a given particle of the medium is destroyed or canceled by the effect of the other pulse. When two pulses with opposite displacements (i.e., one pulse displaced up and the other down) meet at a given location, the upward pull of one pulse is balanced (canceled or destroyed) by the downward pull of the other pulse. Once the two pulses pass through each other, there is still an upward displaced pulse and a downward displaced pulse heading in the same direction that they were heading before the interference. Destructive interference leads to only a momentary condition in which the medium's displacement is less than the displacement of the largest-amplitude wave.

The two interfering waves do not need to have equal amplitudes in opposite directions for destructive interference to occur. For example, a pulse with a maximum displacement of +1 unit could meet a pulse with a maximum displacement of -2 units. The resulting displacement of the medium during complete overlap is -1 unit.

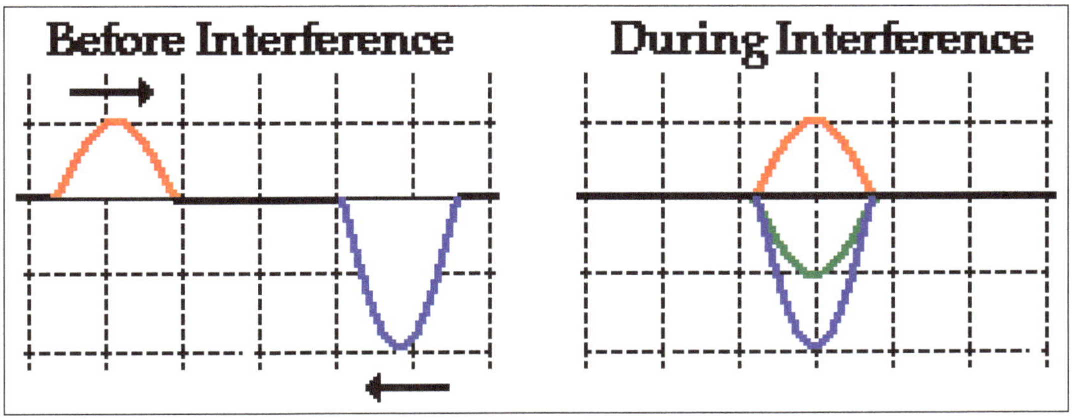

This is still destructive interference since the two interfering pulses have opposite displacements. In this case, the destructive nature of the interference does not lead to complete cancellation.

Interestingly, the meeting of two waves along a medium does not alter the individual waves or even deviate them from their path. This only becomes an astounding behavior when it is compared to what happens when two billiard balls meet or two football players meet. Billiard balls might crash and bounce off each other and football players might crash and come to a stop. Yet two waves will meet, produce a net resulting shape of the medium, and then continue on doing what they were doing before the interference.

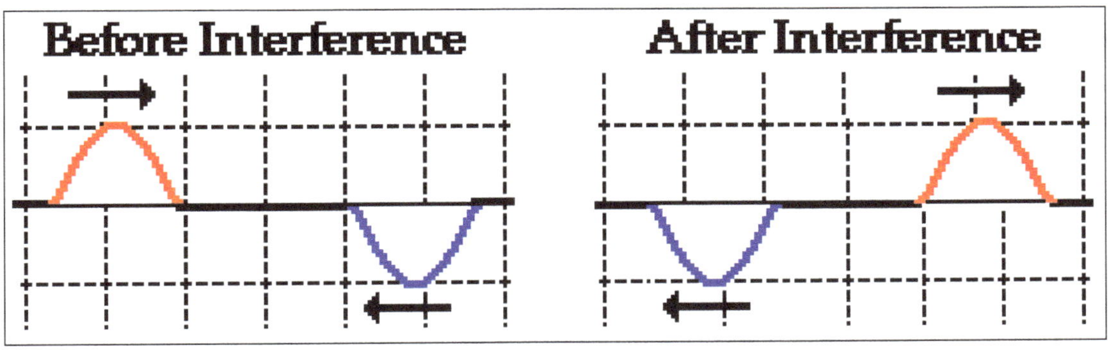

Diffraction

Diffraction of light is defined as the bending of light around corners such that it spreads out and illuminates areas where a shadow is expected. In general, it is hard to separate diffraction from interference since both occur simultaneously. The silver lining which we witness in the sky is caused due to diffraction of light. When the sunlight passes through or encounters the cloud, a silver lining is seen in the sky.

When the double-slit in Young's experiment is replaced by a single narrow slit, a broad pattern with a bright region at the centre is seen. On both sides of the centre, there are alternating dark and bright regions. The intensity becomes weaker away from the centre.

Single Slit Diffraction

In the single-slit diffraction experiment, we can observe the bending phenomenon of light or diffraction that causes light from a coherent source interfere with itself and produce a distinctive pattern on the screen called the diffraction pattern. Diffraction is evident when the sources are small enough that they are relatively the size of the wavelength of light. You can see this effect in the diagram below. For large slits, the spreading out is small and generally unnoticeable.

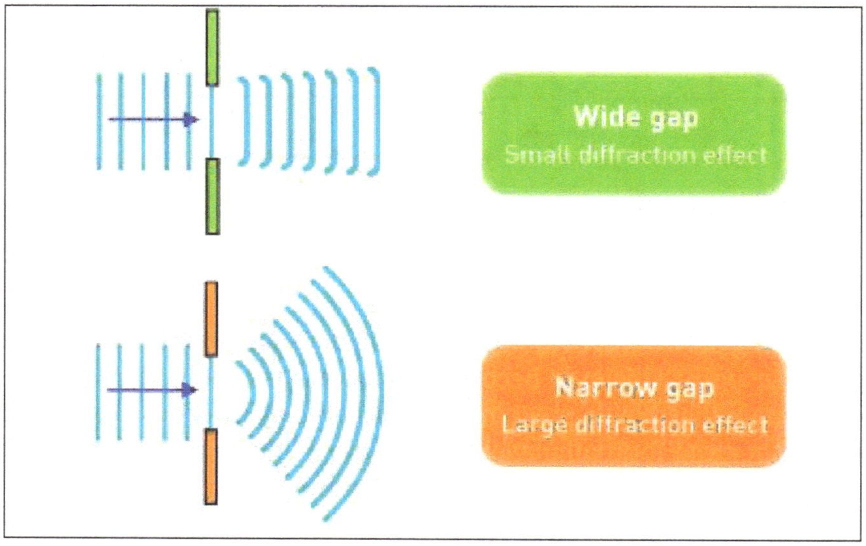

Single Slit Diffraction Formula

We shall assume the slit width a << D. x`D is the separation between slit and source.

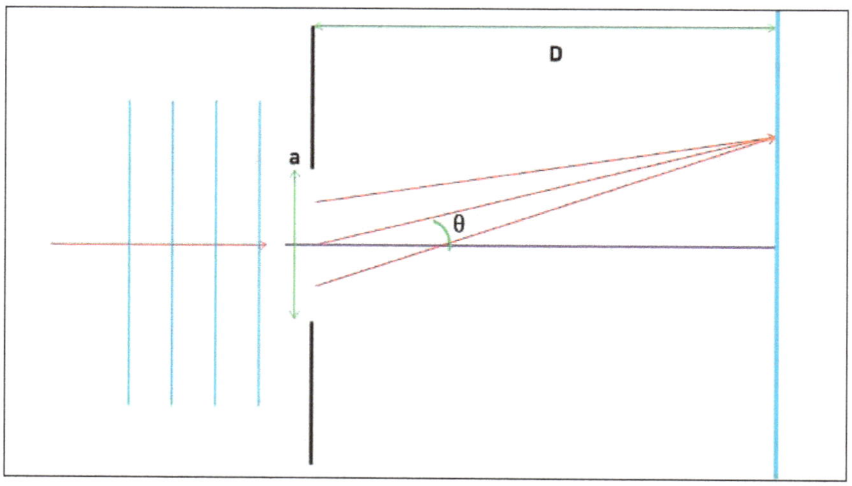

We shall identify the angular position of any point on the screen by ϑ measured from the slit centre which divides the slit by $\frac{a}{2}$ lengths. To describe the pattern, we shall first see the condition for dark fringes. Also, let us divide the slit into zones of equal widths $\frac{a}{2}$. Let us consider a pair of rays that emanate from distances $\frac{a}{2}$ from each other as shown below.

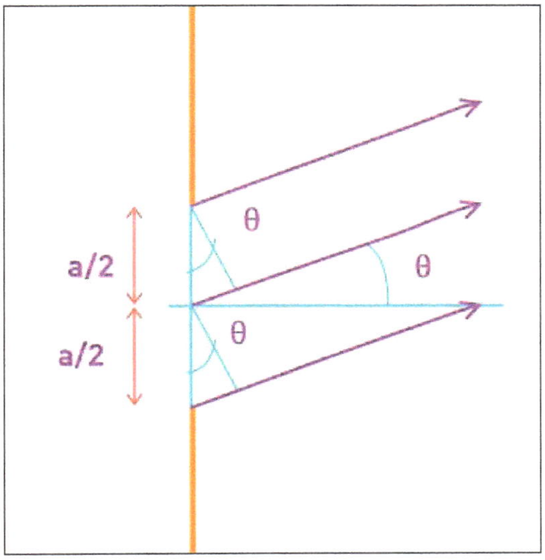

The path difference exhibited by the top two rays shown is:

$$\Delta L = \frac{a}{2}\sin\Theta$$

Remember that this is a calculation valid only if D is very large. We can consider any number of ray pairings that start from a distance $\frac{a}{2}$ from one another such as the bottom two rays in the diagram. Any arbitrary pair of rays at a distance $\frac{a}{2}$ can be considered. We shall see the importance of this trick in a moment.

For a dark fringe, the path difference must cause destructive interference; the path difference must be out of phase by $\frac{\lambda}{2}$. (λ is the wavelength). For the first fringe,

$$\Delta L = \frac{\lambda}{2} = \frac{a}{2}\sin\Theta$$

$$\lambda = a \sin \theta$$

For a ray emanating from any point in the slit, there exists another ray at a distance $\frac{a}{2}$ that can cause destructive interference. Thus, at θ = sin−1λa, there is destructive interference as any ray emanating from a point has a counterpart that causes destructive interference. Hence, a dark fringe is obtained. For the next fringe, we can divide the slit into 4 equal parts of a/4 and apply the same logic. Thus, for the second minima:

$$\frac{\lambda}{2} = \frac{a}{4}\sin\Theta$$

$$2\lambda = a\sin\Theta$$

Similarly, for the nth fringe, we can divide the slit into 2n parts and use this condition as:

$$n\lambda = a \sin \theta$$

The Central Maximum

The maxima lie between the minima and the width of the central maximum is simply the distance between the 1st order minima from the centre of the screen on both sides of the centre. The position of the minima given by y (measured from the centre of the screen) is:

$$\tan\theta \approx \theta \approx y / D$$

For small ϑ,

$\sin \theta \approx \theta$

$$\Rightarrow \lambda = a \sin \theta \approx a\theta$$

$$\Rightarrow \theta = y/D = \lambda a$$

$$\Rightarrow y = \lambda Da$$

The width of the central maximum is simply twice this value:

- Width of central maximum = $2\lambda Da$.
- Angular width of central maximum = $2\theta = 2\lambda a$.

The diffraction pattern and intensity graph is shown below.

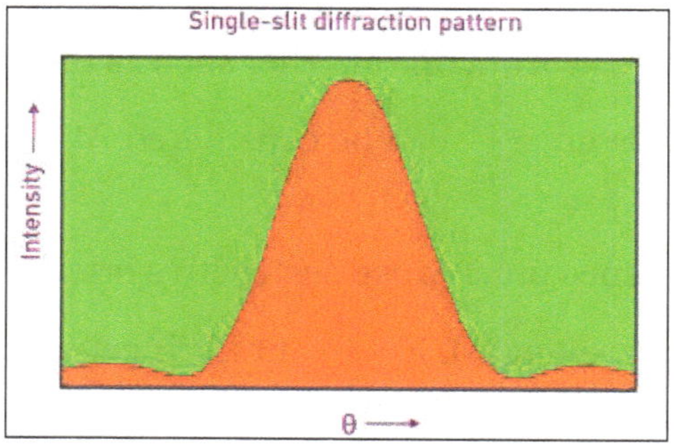

Huygens' Principle

Diffraction refers to what happens to a wave when it hits an obstacle. The key to understanding diffraction is a very simple observation first due to Huygens in 1678. Say a wave arrives at an opaque screen with a little hole in it. On the other side of the screen, the wave equation must still be satisfied with boundary conditions given by the motion of the wave in the hole. That is, the solution is identical to a situation where there was a source in the hole. The wavefront diagram looks like this:

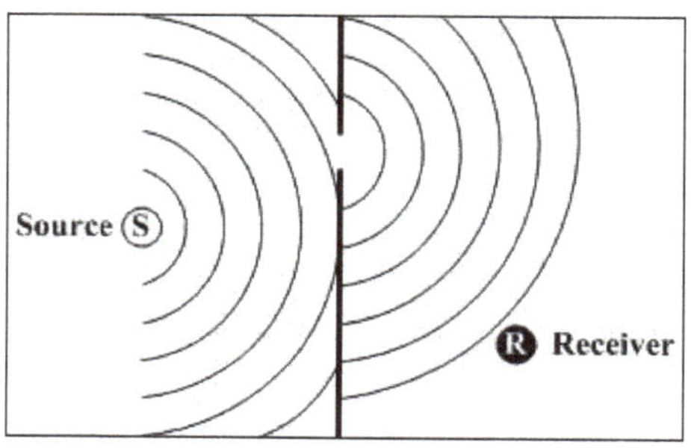

The key point is that the receiver does not care about whatever happened on the other side of the screen. All that matters at is what the wave was like as it passed through the hole. Here is a picture of real wavefronts near a beach in Italy, taken from a satellite, showing this phenomenon:

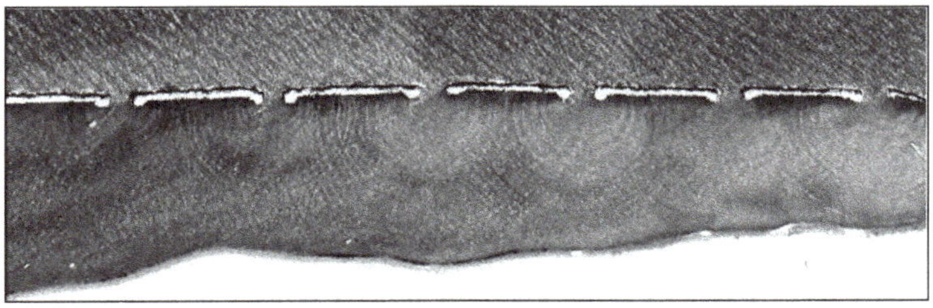

Waves near a shore somewhere in Italy.

The great thing about this way of thinking about diffraction is that, since the wave equation is linear, you can use this trick for any number of holes. You simply add the amplitude for the waves produced from a "source" at each hole:

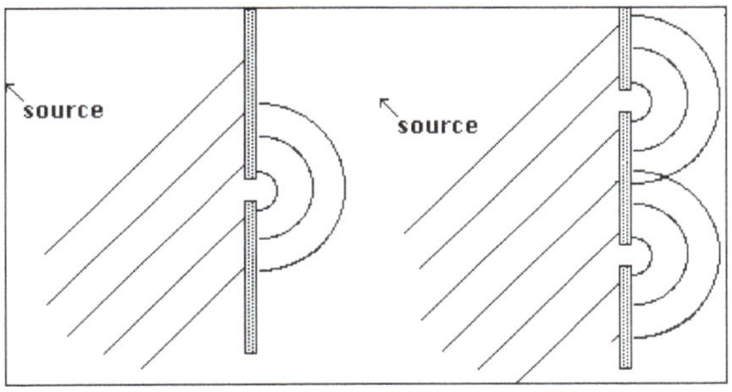

Computing the amplitude by adding point sources in this way is known as Huygens' principle. Huygen's principle works even if the holes are very close together. In fact, it works if they are connected, so instead of a hole, it's a slit. One just has to integrate over the holes rather than sum over them. We'll begin with a discrete set of holes as in a diffraction grating. Then we'll take the continuum limit and talk about slits.

Multiple Hole Diffraction

Using Huygens' principle, we can easily compute the diffraction pattern from a plane wave passing through any number of holes. Say there are N holes in a row separated by a distance d. The solution will be as if there are N sources separated by a distance d. Thus the pattern will be the same as the antenna pattern for this source configuration. Say one source has field $E = E_0 e^{-i\omega t} e^{ikr}$. Recall the diagram for two sources separated by a distance d:

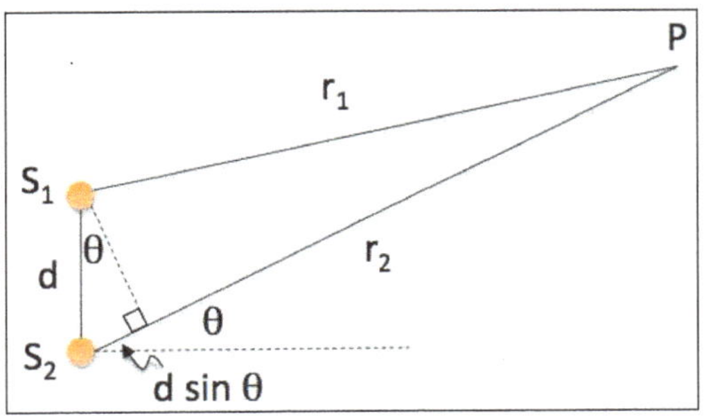

Phase difference from two holes is the same as from two sources.

We found that at a distant point P at an angle θ to the sources the field is:

$$EP = 2E_o e^{-i\omega t} e^{ikr} \cos\left(\frac{\Delta}{2}\right)$$

where,

$$\Delta = 2\pi \frac{d}{\lambda} \sin\theta + \delta$$

with δ the difference in phase between the two sources. This phase difference will be δ = o if the sources are illuminated by a plane coming perpendicular to the separation between the sources. If the wave comes on an angle, δ ≠ o. For N slits, we get N sources, and the picture looks like this,

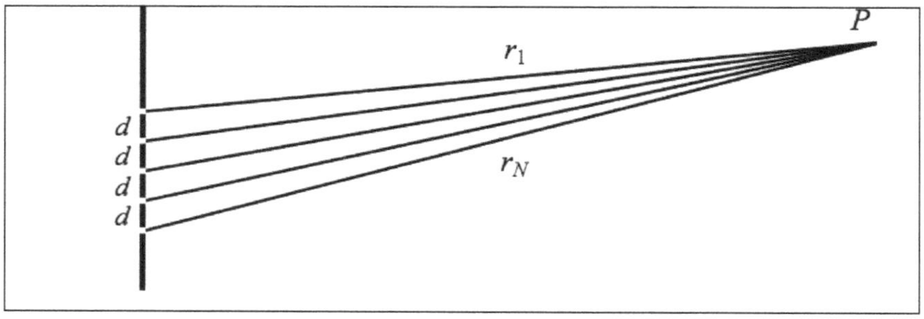

This is called a diffraction grating. In this case, the amplitude at P is:

$$E_p = E_o \left[\frac{e^{iN\frac{\Delta\phi}{2}}}{e^{i\frac{\Delta\phi}{2}}}\right]\left[\frac{\sin\left(N\frac{\Delta}{2}\right)}{\sin\left(\frac{\Delta}{2}\right)}\right]$$

So the intensity is:

$$I = I_o \frac{\sin^2\left(N\frac{\Delta}{2}\right)}{\sin^2\left(\frac{\Delta}{2}\right)}$$

where I_o is the intensity from a single source.

Since the slits are spaced d apart, the total size of the diffraction grating is a = Nd. Now say we place a screen at a distance L from the slits with L ≫ a. Then the height y up this screen is y = L tan θ ≈ L θ ≈ L sin θ, and so,

$$\Delta = 2\pi \frac{d}{\lambda} \frac{y}{L} + \delta$$

If the plane wave hits the grating straight on, so $\delta = 0$ then,

$$I(y) \quad I \frac{\sin\left(N\frac{dy}{} \right)}{\sin\left(\frac{dy}{} \right)}$$

This is the intensity pattern from a diffraction grating. Note that for N = 2 this simplifies to $I(y) = 4I_0 \cos\left(\pi \frac{dy}{\lambda L} \right)$. Here are some plots of I(y) for N = 2, N = 5 and N = 10 slits.

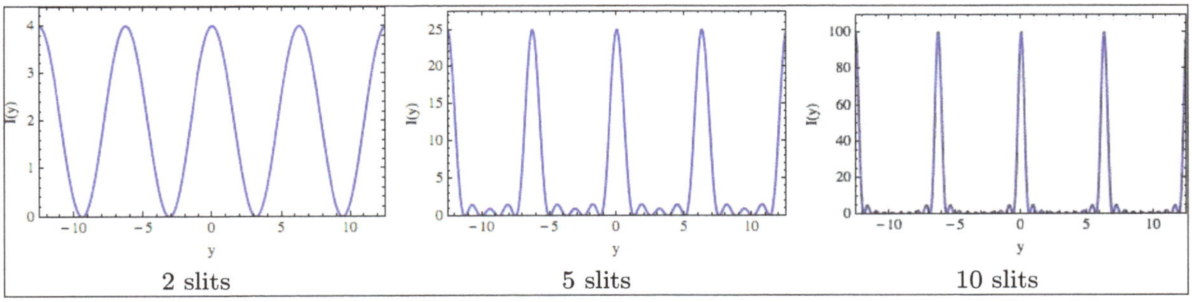

Patterns from diffraction gratings with N = 2, 5 and 10 slits. These plots show the intensity at a distance y up a screen placed at a distance L from the wall.

Here's what this actual looks like with a laser passing through a diffraction grating.

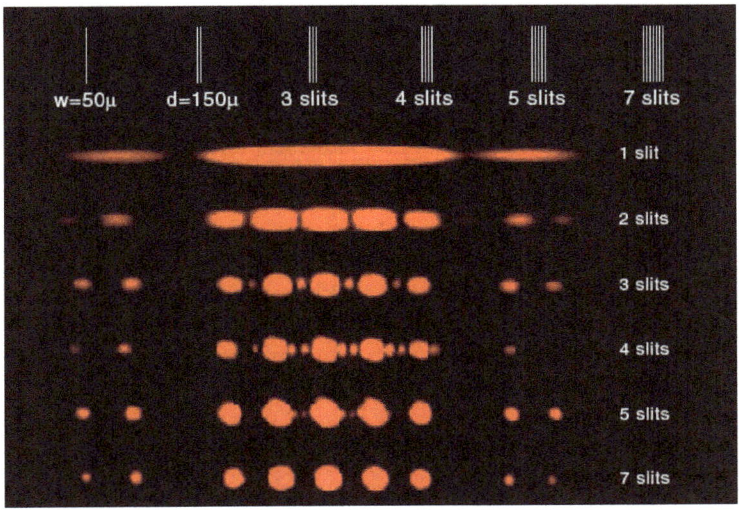

Pattern from a diffraction grating. As the number of slits is increased, the peaks become clearly separated. The extra dots on the right and left side after the gaps are due to the finite width of the slits.

You can see from the plots that each intensity pattern has a maximum at y = 0. Indeed, we find:

$$I_0 = \lim_{y \to 0} I(y) = I_0 N^2$$

which agrees with the vertical axes on the plots. So the more slits in the grating, the larger the central peak will be. The small bumps are due to the oscillation of the $\sin^2\left(\pi N\frac{yd}{\lambda L}\right)$ factor in the numerator. Indeed, the factor of N makes these oscillations have very high frequency. The overall periodicity is due to the denominator factor $\sin^2\left(\pi\frac{dy}{\lambda L}\right)$. This vanishes when,

$$y_{max} = j\frac{L\lambda}{d} = j\frac{L}{d}\frac{c}{v}, \quad j = 0,1,2,\ldots$$

with v the frequency.

Since the peaks are spaced by $\frac{L}{d}\lambda$, different wavelengths will have peaks in different places. Thus, diffraction gratings can be used to characterize the spectra of various things. For example, gases have interesting spectra which can be resolved with diffraction gratings.

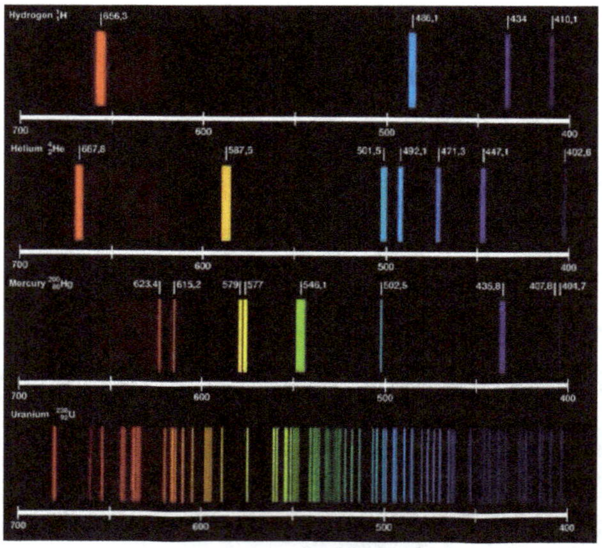

Spectra of hydrogen, helium, mercury and uranium as viewed through a diffraction grating.

The wavelengths in the hydrogen spectrum have the interesting pattern $\frac{1}{\lambda} = \frac{1}{91\text{nm}}\left(\frac{1}{4} - \frac{1}{j^2}\right)$ for j = 3, 4, 5 and 6. This phenomenological observation was known in the 19th century.

Resolving Power

An important use of diffraction gratings is to separate light into different wavelengths, as in Figure. Say a gas emits two colors, with wavelengths λ and λ′. How do we know if a given diffraction grating can separate the colors? Obviously, if the two wavelengths are infinitesimally close together, we will not be able to separate them. A reasonable criterion for whether frequencies can be separated is if the maximum of the intensity pattern from one wavelengths lies on top of the minimum of the intensity pattern of the other:

$$\max \lambda' = \min \lambda$$

This is known as the Rayleigh Criterion. All frequencies have maxima at y = 0, so we must look to the second peak to apply the criterion. The second maximum for λ' is at,

$$y_{max} = \frac{d}{L}\lambda'$$

The minima for λ happen only when the $\sin^2\left(\pi N\frac{yd}{\lambda L}\right)$ numerator in Equation $I(y) = I_0 \frac{\sin^2\left(\pi N\frac{dy}{\lambda L}\right)}{\sin^2\left(\pi\frac{dy}{\lambda L}\right)}$ vanishes, which is when,

$$y_{min} = m\frac{L\lambda}{dN},\ m = 1,\ 2,\ \cdots,\ N-1,\ N+1,\ldots$$

Values m = 0 and m = N are excluded from this list since for these values the dominator in Equation ($I(y) = I_0 \frac{\sin^2\left(\pi N\frac{dy}{\lambda L}\right)}{\sin^2\left(\pi\frac{dy}{\lambda L}\right)}$) vanishes as well and they are associated with maxima not minima.

For the second maximum of λ' to overlap with a minimum of λ we then want,

$$\frac{L}{d}\lambda' = m\frac{L}{d}\frac{\lambda}{N}$$

or,

$$\frac{\lambda'}{\lambda} = \frac{m}{N}$$

Note that this is independent of L and d. We want the value of m where this ratio is as close to 1 as possible (so that λ' is as close to λ as possible). So we take m = N – 1, referring to the minimum right next to the first maximum. Then $\frac{\lambda'}{\lambda} = \frac{N-1}{N} = 1 - \frac{1}{N}$. Thus $\frac{1}{N} = 1 - \frac{\lambda'}{\lambda} = \frac{\Delta\lambda}{\lambda}$ or,

$$\frac{\lambda}{\Delta\lambda} = N$$

where $\Delta\lambda = \lambda' - \lambda$. In general $\frac{\lambda}{\Delta\lambda}$ with $\Delta\lambda$ is the smallest shift in λ which can be resolved is known as the resolving power of an optical system. Equation above gives the resolving power of a diffraction grating.

For example, Hydrogen has wavelengths λ_{red} = 656.2 nm, λ_{blue} = 486.1 nm, $\lambda_{violet-1}$ = 434.0 nm and $\lambda_{violet-2}$ = 410.1 nm. As you can see from Figure the closest are 434.0 nm and 410nm with,

$$\frac{410}{434-410} = 18$$

So we would need 18 slits to see this separation. With 18 slits the lines would be barely separable. Typical diffraction gratings have 100s slits, so separations of $\frac{\Delta\lambda}{\lambda} \sim 1\%$ can be resolved. High quality diffraction gratings can have ~10, 000 slits.

Wide Slits

What happens if instead of having N holes separated d apart, we have one big hole? An easy way to derive the result is to take the limit of our previous result, Eq. ($I(y)=I_o \dfrac{\sin^2\left(\pi N \frac{dy}{\lambda L}\right)}{\sin^2\left(\pi \frac{dy}{\lambda L}\right)}$) $N \to \infty$ holding the overall size a = Nd fixed. Writing Eq. ($I(y)=I_o \dfrac{\sin^2\left(\pi N \frac{dy}{\lambda L}\right)}{\sin^2\left(\pi \frac{dy}{\lambda L}\right)}$) in terms of N and I we have,

$$I(y)=I_o \frac{\sin^2\left(\pi \frac{Ndy}{\lambda L}\right)}{\sin^2\left(\pi \frac{dy}{\lambda L}\right)}$$

Note that $I(0) = \lim_{y\to 0} (y) = I_0 N^2$ which seems to blow up as $N \to \infty$. However, the relative intensity at a distance y compared to the intensity at y = 0 is finite,

$$\frac{I(y)}{I(0)} = \lim_{N\to\infty} \frac{1}{N^2} \frac{\sin^2\left(\pi \frac{Ndy}{\lambda L}\right)}{\sin^2\left(\pi \frac{dy}{\lambda L}\right)} = \lim_{N\to\infty} \frac{1}{N^2} \frac{\sin^2\left(\pi \frac{ay}{\lambda L}\right)}{\sin^2\left(\pi \frac{ay}{N\lambda L}\right)} = \left(\frac{\sin\left(\pi \frac{ay}{\lambda L}\right)}{\pi \frac{ay}{\lambda L}}\right)^2$$

This function is known as the sinc function,

$$\text{sinc}(\beta) \equiv \frac{\sin\beta}{\beta}$$

So,

$$\frac{I(y)}{I(0)} = \text{sin c}^2\left(\pi \frac{ay}{\lambda L}\right)$$

where a is the size of the slit, λ the wavelength, L the distance to the screen and y the distance up the screen. This pattern looks like:

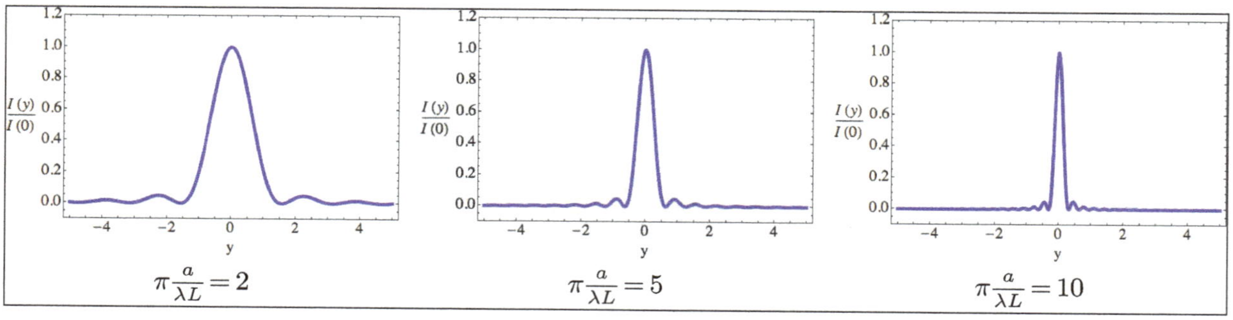

The $\operatorname{sinc}^2\left(\frac{\pi a}{\lambda L}y\right)$ function gives the intensity from a wide slit. The wider the slit, the narrower the peak at fixed λ.

Or we can draw the picture like this,

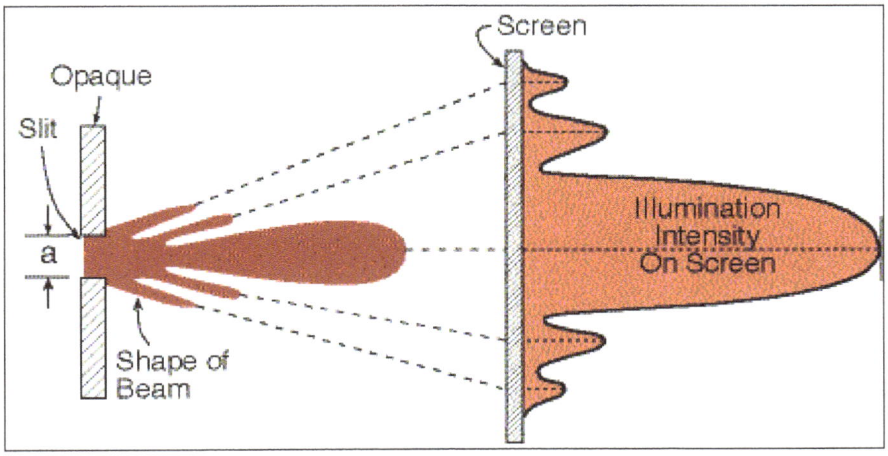

Diagram of wide-slit diffraction pattern.

The location of the first minimum in $\operatorname{sinc}^2(\beta)$ is at $\beta = \pi$ or $\pi\frac{ay}{\lambda L} = \pi$, thus the width of the first peak is,

$$\Delta y = \frac{2\lambda L}{a}$$

Note that as a gets smaller, y gets bigger. In the limit that a goes to a point, the outgoing wave is in phase everywhere and $\Delta y \to \infty$. Here's a photo of what this actually looks like:

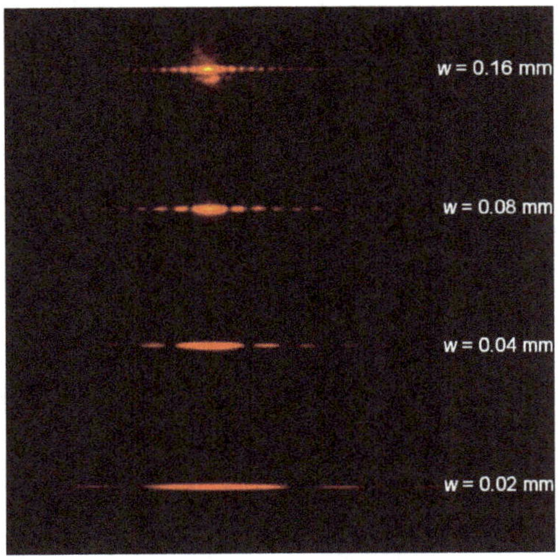

Wide slit interference patterns with variable width slits.

Now say we have two far away sources (like stars) and the light passes through a slit, making an image on a screen. What angular separation of the stars will we be able to distinguish on the screen? To find out, we apply the Rayleigh criterion: the maximum of one should hit the minimum of the other. For light coming in at an angle $\Delta\theta$, its plane-wave light will have slightly different phases at different parts of the slit. The net effect is to shift the pattern so that the maximum is not at y = 0 but at y = L sin θ. So,

$$\frac{I(y)}{I(0)} = \operatorname{sinc}^2\left[\pi\frac{a}{\lambda L}(y - L\sin\theta)\right]$$

Say one star's light comes in at θ = 0 and the other at Δθ. The one at θ = 0 is directly normal to the screen, so the minimum of its intensity pattern is at $y = \frac{\lambda L}{a}$. If the other one comes it at an angle Δθ, so its maximum is at $y = L\sin\Delta\theta \approx L\,\Delta\theta$ for small angles. Setting these equal gives,

$$\Delta\theta_R = \frac{\lambda}{a}$$

This is the resolving power of a slit. For a circular aperture, the formula is:

$$\Delta\theta_R = 1.22\frac{\lambda}{a}$$

where the 1.22 comes from the circular geometry. If you have 20/20 vision, you can only resolve objects at around 12 times farther apart than the Rayleigh criterion. That is, ordinary vision is not diffraction limited.

If we have two wide slits, the interference pattern is a combination of the $\operatorname{sinc}^2\left(\pi\frac{ay}{\lambda L}\right)$ behavior of a wide slit and the $\cos^2\left(\pi\frac{dy}{\lambda L}\right)$ behavior of the double slit. This is shown in Figure below.

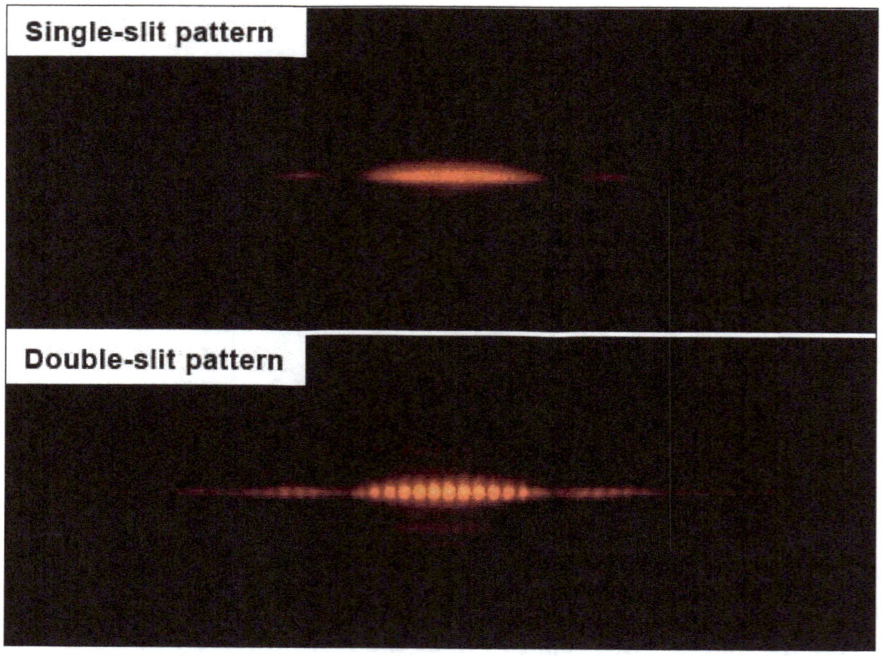

Comparison of the interference pattern for a single wide slit and two wide slits.

Fourier Optics

What would happen if instead of a slit, we put a film in the slit which had a transparency T(x) which varies along the length of the slit. Let's define T = 0 as opaque and T = 1 as totally transparent. Thus, the wide slit is T(x) = 1. A half transparent slit would have T(x) = 1/2. The diffraction grating is a bunch of small regions with T = 1 surrounded by regions with T = 0. We can model this with a set of δ-functions:

$$T(x)=\sum_{j=1}^{N}\delta(x-jd)$$

What is the intensity pattern on the screen from a general T(x)? We can compute the answer using Huygens' principle.

If a plane wave comes in with a field E_0 at the slit, then each point will come with a field given by $E_0T(x)$. At a distance y on a screen, we can then use our formula that the phase shift is:

$$\Delta=2\pi\frac{x}{\lambda}\frac{y}{L}$$

So the sum of the phases gives,

$$E(y)\propto\int dxT(x)e^{2\pi ix\frac{y}{L\lambda}}=\overline{T}\left(2\pi\frac{y}{L\lambda}\right)$$

where,

$$\overline{T}(k)=\int dxT(x)e^{ikx}$$

is the Fourier transform of k. That is, the image on the screen is the Fourier transform of the transparency. So finally, we understand Fourier optics. You have been using this result in lab all semester, and we have finally derived it.

To make sure we're not nuts, let's check on our wide slit. There we expect the field E to be the square root of the intensity in Eq.($\frac{I(y)}{I(0)}=\operatorname{sinc}^2\left(\pi\frac{ay}{\lambda L}\right)$)

$$E(y)\propto\operatorname{sinc}\left(\pi\frac{ay}{\lambda L}\right)$$

The relevant T (x) is given in Eq. $\frac{I(y)}{I(0)}=\operatorname{sinc}^2\left[\pi\frac{a}{\lambda L}(y-L\sin\theta)\right]$. Its Fourier transform is,

$$\overline{T}(k)=\int_{-\infty}^{\infty}dxT(x)e^{ikx}=\int_{-\frac{a}{2}}^{\frac{a}{2}}dxe^{ikx}=\frac{1}{ik}\left[e^{i\frac{ka}{2}}-e^{-i\frac{ka}{2}}\right]=\frac{2}{k}\sin\left(\frac{ka}{2}\right)=a\operatorname{sinc}\left(\frac{ka}{2}\right)$$

So,

$$E(y)\propto\overline{T}\left(\pi\frac{ay}{L\lambda}\right)$$

exactly as expected. We can also check the diffraction grating using Eq. ($T(x)=\sum_{j=1}^{N}\delta(x-jd)$),

$$\overline{T}(k)=\int_{-\infty}^{\infty}dxT(x)e^{ikx}=\sum_{j=1}^{N}\int_{-\infty}^{\infty}dx\delta(x-jd)e^{ikx}=\sum_{j=1}^{N}e^{ikjd}=e^{i(N+1)}\Delta\frac{\sin\left(N\frac{\Delta}{2}\right)}{\sin\left(\frac{\Delta}{2}\right)}$$

where $\Delta = kd = 2\pi\frac{d}{\lambda}$. This agrees with our result for the diffraction grating in Eq. $E_p=E_o\left[\frac{e^{iN\frac{\Delta\phi}{2}}}{e^{i\frac{\Delta\phi}{2}}}\right]\left[\frac{\sin\left(N\frac{\Delta}{2}\right)}{\sin\left(\frac{\Delta}{2}\right)}\right]$.

In summary, if we put a screen over the slit with transparency T(x) then the image on the screen will be proportional to $\overline{T}\left(\pi\frac{ay}{L\lambda}\right)$. Note that this holds even for monochromatic (fixed frequency) incoming light, as in a laser. Say the wavelength λ of light and the distance L to the screen are fixed. Then the intensity at a distance y up the screen tells us the strength of the Fourier component of the slit corresponding to a wavenumber of,

$$k=2\pi\frac{y}{L\lambda}$$

To be absolutely clear, this k is not $\frac{2\pi}{\lambda}$. The Fourier transform on the screen has nothing to do with the wavelength of the light entering the slit. The farther up the slit you look, the larger k, and hence smaller wavelength modes you are seeing. This is why the spot gets smaller as the slit gets bigger.

Dispersion

In optics, dispersion refers to light that is separated by individual colors as it passes through an object. Rainbows are the most common example of this event. The rainbow forms when white light is separated into components of different wavelengths (or colors). Dispersion can result in signal degradation in many applications, especially over large distances.

Dispersion in Optical Fibers

Chromatic Dispersion (CD)

Light within a medium travels at a slower speed than in vacuum. The speed at which light travels is determined by the medium's refractive index. In an ideal situation, the refractive index would not depend on the wavelength of the light. Since this is not the case, different wavelengths travel at different speeds within an optical fiber.

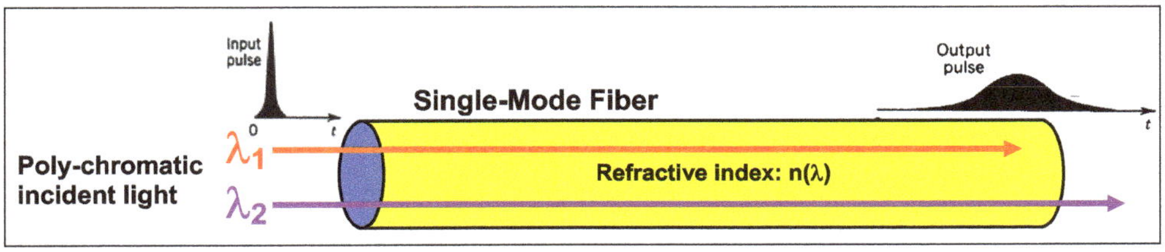

CD in single-mode fiber.

Laser sources are spectrally thin, but not monochromatic. This means that the input pulse contains several wavelength components, traveling at different speeds, causing the pulse to spread. The detrimental effects of chromatic dispersion result in the slower wavelengths of one pulse intermixing with the faster wavelengths of an adjacent pulse, causing intersymbol interference. The Chromatic Dispersion of a fiber is expressed in ps/(nm*km), representing the differential delay, or time spreading (in ps), for a source with a spectral width of 1 nm traveling on 1 km of the fiber. It depends on the fiber type, and it limits the bit rate or the transmission distance for a good quality of service.

CD of Current Network Spans

As a consequence of its optical characteristics, the Chromatic Dispersion of a fiber can be changed by acting on the physical properties of the material. To reduce fiber dispersion, new types of fiber were invented, including dispersion-shifted fibers (ITU G.653) and non-zero dispersion-shifted fiber (ITU G.655). The most commonly deployed fiber in networks (ITU G.652), called "dispersion-unshifted" single-mode fiber, has a small chromatic dispersion in the optical window around 1310 nm, but exhibits a higher CD in the 1550 nm region. This dispersion limits the possible transmission length without compensation on OC-768/STM-256 DWDM networks.

ITU G.653 is a dispersion-shifted fiber (DSF), designed to minimize chromatic dispersion in the 1550 nm window with zero dispersion between 1525 nm and 1575 nm. But this type of fiber has several drawbacks, such as higher polarization mode dispersion than ITU G.652, and a high Four Wave Mixing risk, rendering DWDM practically impossible. For these reasons, another single-mode fiber was developed: the Non-Zero Dispersion-Shifted Fiber (NZDSF). NZDSF fibers have now replaced DSF fibers, which are not used anymore. The ITU G.655 Non-Zero Dispersion-Shifted Fibers were developed to eliminate non-linear effects experienced on DSF fibers. They were developed especially for DWDM applications in the 1550 nm window. They have a cutoff wavelength around 1310 nm, limiting their operation around this wavelength.

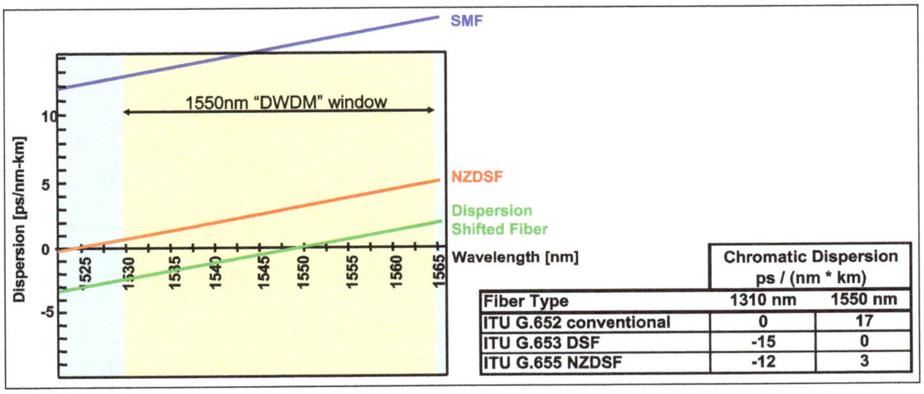

Fiber Type	Chromatic Dispersion ps / (nm * km) 1310 nm	1550 nm
ITU G.652 conventional	0	17
ITU G.653 DSF	-15	0
ITU G.655 NZDSF	-12	3

Chromatic dispersion profiles of different fiber types.

CD Limit and Compensation

The chromatic dispersion in fiber causes a pulse broadening and degrades the transmission quality, limiting the distance a digital signal can travel before needing regeneration or compensation. For DWDM systems using DFB lasers, the maximum length of a link before being affected by chromatic dispersion is commonly calculated with the following equation:

$$L = \frac{104{,}000}{CD * B^2}$$

L is the link distance in km, CD is the chromatic dispersion in ps/(nm*km), and B is the bit rate in Gbps.

As an example, consider a typical network transmitting data at 10 Gbps on a channel at 1550 nm over a standard ITU G.652 fiber, having a CD coefficient of 17 ps/(nm*km). In this case, the theoretical maximum distance of the link, before adding CD compensators, will be around 61 km, which is close to the typical node distance in Europe. This length will be divided by 16 when upgrading the network to a 40 Gbps bit rate, thus falling under 4 km. This distance is around 20 km for a 40 Gbps transmission through NZDSF. We can thus see that CD is a major limiting factor in high-speed transmission.

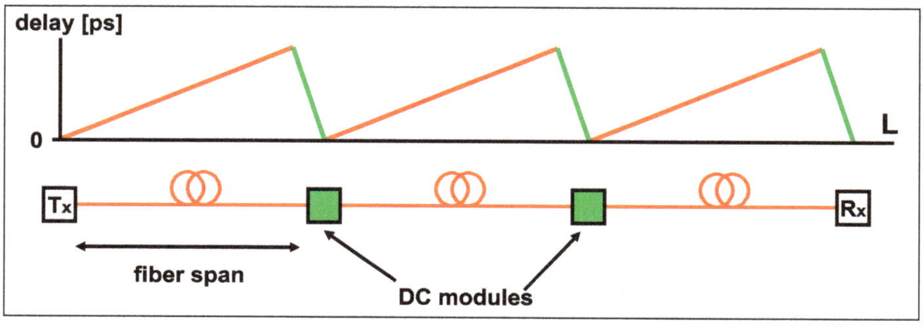

Chromatic dispersion compensation scheme.

Fortunately, CD is quite stable, predictable, and controllable. Dispersion Compensation Fiber (DCF), with its large negative CD coefficient, can be inserted into the link at regular intervals to minimize its global chromatic dispersion. While each spool of DCF adequately solves chromatic dispersion for one channel, this is not usually the case for all channels on a DWDM link. At the extreme wavelengths of a band, dispersion still accumulates and can be a significant problem. In this case, a tunable compensation module may be necessary at the receiver.

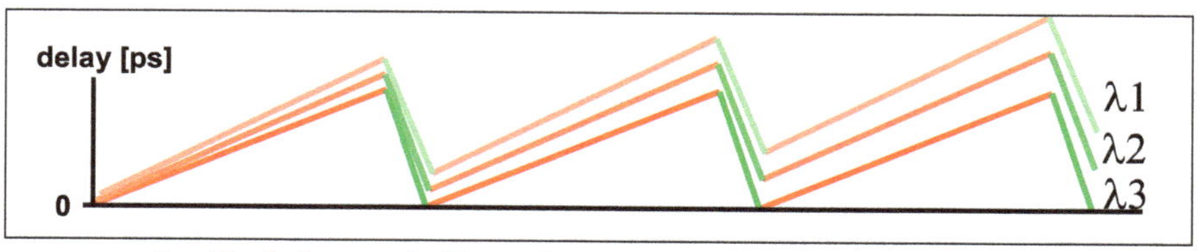

Residual chromatic dispersion on a DWDM-compensated link.

Therefore, chromatic dispersion measurement is essential in the field to verify the types of installed fibers. Such measurements assess if and how the fibers can be upgraded to transmit higher bit rates, verify fiber zero point and slope for new installations, and carefully evaluate compensation plans.

CD Measurement Methods

In the field, there are three main methods for determining the chromatic dispersion of an optical fiber. These are described by three TIA/EIA industry standards: the pulse-delay method (FOTP-168 standard), the modulated phase-shift method (FOTP-169 standard), and the differential phase shift method (FOTP-175 standard).

These methods all measure first the time delay, in ps, as a function of the wavelength. They then deduce the chromatic dispersion coefficient, in ps/(nm*km), from the slope of this delay curve and from the length of the link. The delay curve plotted by CD analyzers is a fitted trace based on the acquired time delays at discrete wavelength points. Therefore, the accuracy of the computation of the CD coefficient depends on the amount of the measured data.

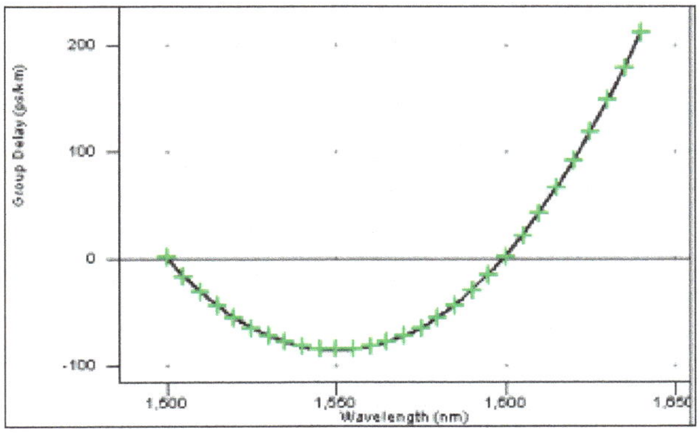

Curve fit to calculate CD coefficient.

Phase-shift and differential phase-shift methods are quite similar. In both methods, a modulated source is injected at the input of the fiber under test. The phase of the sinusoidal modulating signal is analyzed at the output of the fiber and compared to the phase of a reference signal, modulated with the same frequency. In the phase-shift method, the reference signal has a fixed wavelength, while the other modulated signal is tuned in wavelengths. In the differential phase-shift method, both signals are tuned in wavelengths, with a fixed wavelength interval. The analyzed modulated signal, tuned in wavelengths, is compared to a close reference signal, also tuned in wavelength, but the wavelength gap is constant. The time delay of the link is deduced from the phase-shift measurement, using the relationship between the delay (t), the phase (φ), and the modulation frequency (f):

$$\Delta t = \frac{\Delta \phi}{2\pi f}$$

The phase-shift methods assume there is access to both ends of the fiber under test, with a transmitter unit connected at the input, and a receiver unit at the output of the fiber. The wavelength selection can be achieved either at the transmitter unit level, using a tunable laser, or at the receiver unit level, with a broadband source at the input and a wavelength tunable filter in the receiver unit. This filter must be spectrally thin (FWHM < 1 nm) to ensure low impact on the measurement accuracy. This second alternative is commonly used in the differential phase-shift method. One disadvantage of this method is that the phase-shift is evaluated comparing two signals with close wavelengths. Increasing the wavelength interval between the two modulated signals will increase the accuracy of the delay at one individual wavelength point, but will decrease the number data points that can be acquired to trace the delay curve, leading to a lower accuracy in the CD coefficient computation. A compromise has to be found between the wavelength interval for the differential phase-shift calculation and the number of acquired points to fit the delay curve, but achieving this is not straightforward.

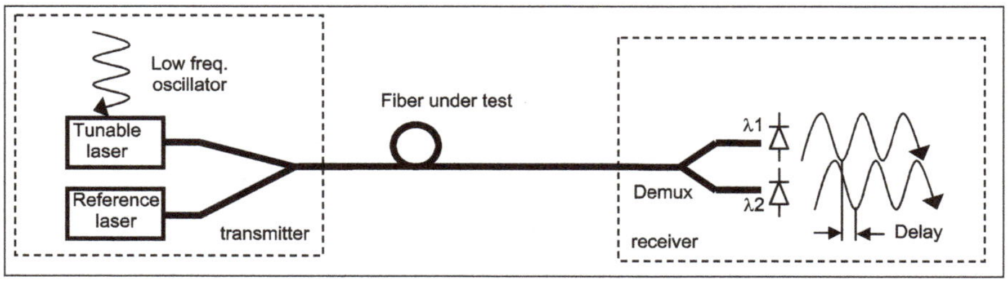

Phase-shift principle.

The phase-shift methods are two-ended solutions, allowing measurement through amplifiers on long-haul links. They measure the CD with high accuracy, since the wavelengths of modulated signals are known accurately, and many points of measurements can be acquired, leading to a better fit of the delay curve. However, these solutions are expensive and thus may not be appropriate for metropolitan applications. They require operators at both ends of the link, often with bulky instruments.

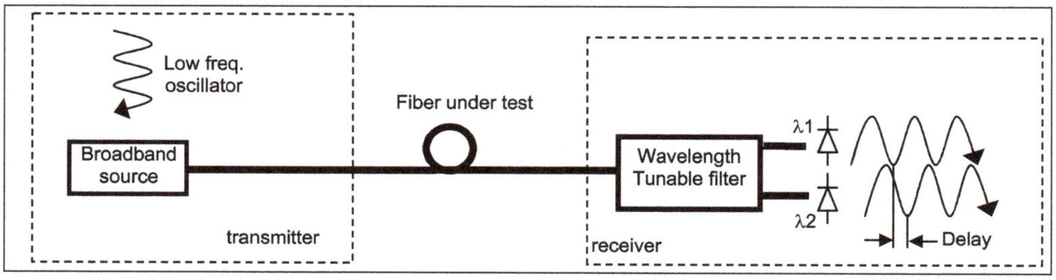

Differential phase-shift principle.

The pulse-delay methods measure the time that different wavelength carriers travel through the fiber under test, either by photon counting (a direct but very complex method), or by measuring the link length with a multi-wavelength OTDR. The CD-OTDR launches multiple laser pulses into one end of the fiber under test, ideally using more than four different wavelengths for better accuracy. It then analyzes the time to return after a back-reflection from the connector at the other end. The time delay as a function of the wavelength is deduced by comparing the times of flight of the laser pulses.

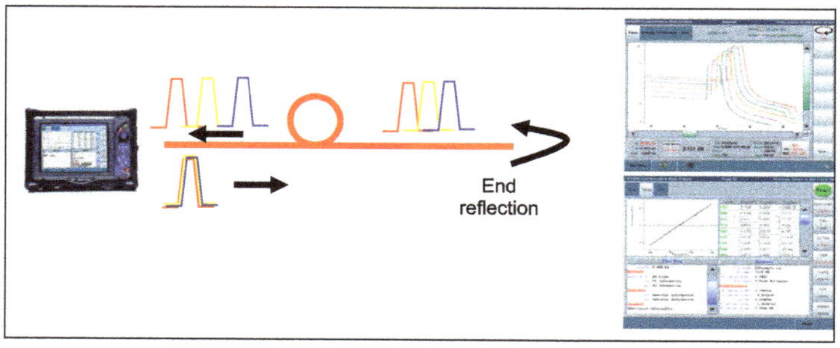

CD-OTDR principle.

The CD-OTDR solution is highly cost-effective, primarily because the initial cost of the instrument is far lower than that of a phase-shift setup. Furthermore, the CD-OTDR is also a complete 3-wavelength OTDR module, which is at the heart of any fiber characterization effort. The CD-OTDR combines two test setups in a single, small instrument. It is a one-ended test solution, resulting in less manpower to carry out the test, less training cost, and lower transportation costs. It can be less accurate than a phase-shift tester, but accuracy better than 5% can be achieved with a 6-wavelength CD-OTDR. This is sufficient to assess a link's chromatic dispersion, which is a stable and predictable parameter. The CD-OTDR is therefore a good alternative to phase-shift testing in metro rings and regional networks.

Table: Comparison table of CD measurement methods.

	Phase-shift method	CD-OTDR method
Application	Long-haul link	Metro & access link
Measurement constraint	Two-ended measurement	One-ended measurement
Accuracy	Good accuracy (depending on Number of acquired points)	<5% with 6-wavelength CD-OTDR
Cost	High	Low + 3-wavelength OTDR testing capability

Polarization Mode Dispersion (PMD)

Testing for Polarization Mode Dispersion (PMD) is becoming essential before upgrading a network to a higher bit rate because PMD can highly degrade the quality of transmission. It is a difficult parameter to measure, however, because it varies with time and depends on environmental conditions. PMD is known to stem from the difference in the propagation constants of a fiber due to geometrical imperfections in the fiber. The term "PMD" denotes both the physical phenomenon and the associated temporal delay. PMD also causes a system penalty because of the associated pulse spreading in a high-speed digital transmission system.

The physical origin of PMD is essentially linear birefringence due to core eccentricity and ovalization. These appear during the manufacturing process or result from external stresses on the fiber, such as bends and twists, and can be considered constant over a length called the "coupling length". The typical value of the coupling length is several hundred meters and depends on fiber manufacturing parameters. This means that for distances that are practical for transmission applications, the actual length of the fiber is much greater than the coupling length.

The PMD phenomenon is characterized by Differential Group Delay (DGD). DGD is the difference in propagation time between the two polarization eigenstates, which are the states of polarization with minimum and maximum propagation time for each wavelength.

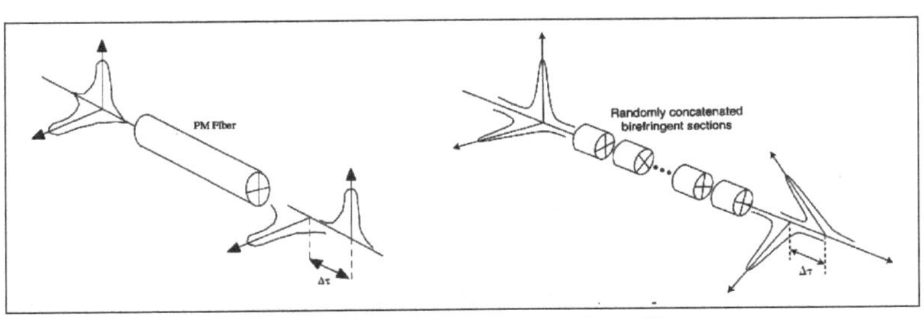

DGD of PM and random coupling fibers.

In case of weak mode coupling (Polarization Maintaining Fiber-short length of ordinary fiber), the light polarized along the slow axis arrives later than the light traveling along the fast axis (i.e., the fast and slow axes have different indexes of refraction). In this case, the PMD is equal to the DGD. In other cases (long fiber lengths), the optical fiber acts like many short birefringent elements stacked together and the alignment of fast- and slow-axes is random from element to element. Consequently, we speak about random (or strong) mode coupling. In that case, the DGD varies as a function of wavelength and the PMD, expressed in ps, is the average value of the DGD spectral distribution.

The average DGD scales as the square root of the length of the fiber. So the PMD coefficient, expressed in ps/√km, is often calculated. In addition, the second-order PMD coefficient, in ps/(nm. km), expresses the PMD dependency with the wavelength. PMD needs to be tested on the C&L bands. But, depending on the wavelength transmission window of the network, there is a need to also test PMD at 1310 nm as PMD values could be different from 1310 nm to 1550 nm.

The Statistical Nature of PMD

For a practical transmission system, DGD determines the system penalty and depends to a large extent on the wavelength of operation within the operating wavelength band. But DGD also changes with environmental conditions over time. The next two traces show DGD as a function of wavelength, for the same fiber at different times.

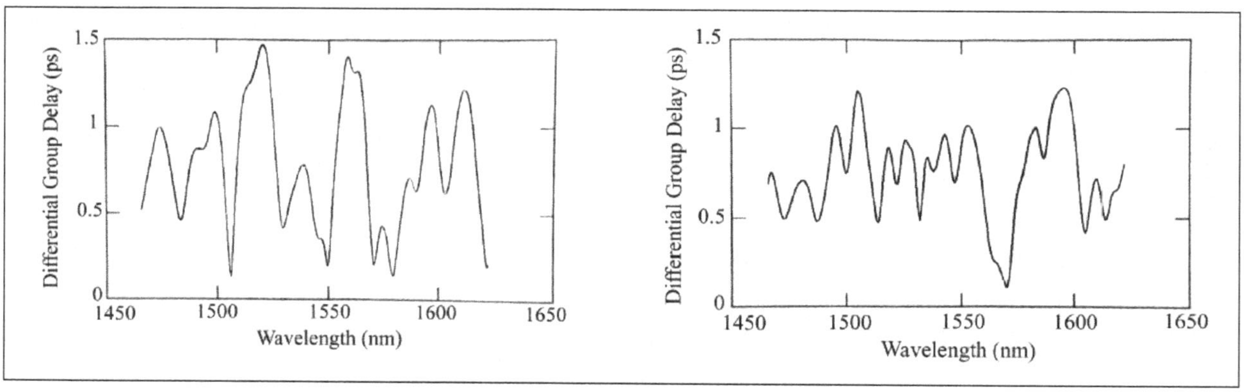

DGD versus wavelength, on the same fiber, at two different moments.

The graphs show that DGD at a particular wavelength changes with time. The variation can be as much as 1 ps in a few minutes. The general aspect of the plots is nevertheless the same, and the distribution statistics remain constant. It can be shown that DGD versus wavelength exhibits a Maxwellian distribution, with a fairly constant mean value over time. The PMD figure is usually taken as the average of the wavelength distribution. For this fiber, the PMD (average value of DGD) is 0.65 ps.

PMD Limit and Compensation

PMD is inevitable because it is caused by fiber stresses during manufacturing or by environmental condition changes. But a small PMD level can be tolerated in networks, depending on the data rate that is intended to be transmitted. Generally, the maximum tolerable PMD is typically 10% of the bit time.

Table: Bit rate and time, and PMD limit.

SDH	SONET	Transmission Rate	Bit Time	PMD Limit
	OC-1	51.84 Mb/s	19.29 ns	2 ns
STM-1	OC-3	155.52 Mb/s	6.43 ns	640 ps
STM-4	OC-12	622.08 Mb/s	1.61 ns	160 ps
	OC-24	1,244.16 Mb/s (1.2 Gbps)	803.76 ps	80 ps
STM-16	OC-48	2,488.32 Mb/s (2.5 Gbps)	401.88 ps	40 ps
STM-64	OC-192	9,953.28 Mb/s (10 Gbps)	100.47 ps	10 ps
STM-256	OC-768	39,318.12 Mb/s (40 Gbps)	25.12 ps	2.5 ps

Since PMD is an unstable phenomenon, it is not easy to compensate for it. However, Dispersion Compensation Modules (DCM) has been developed for that purpose. DCMs are placed in front of receivers on the network and can be tuned in dispersion to compensate for the dispersion measured continuously on a sample of optical pulses. Regularly measuring PMD in the field is essential in order to evaluate network capacity and assess the possibility of upgrading networks for higher bit rate transmission.

PMD Measurement Methods

In the field, there are three main methods for determining the polarization mode dispersion of an optical fiber, described by three TIA/EIA industry standards: the Fixed Analyzer Method (FOTP-113 standard), the Jones Matrix Method (FOTP-122 standard), and the Interferometric Method (FOTP-124 standard).

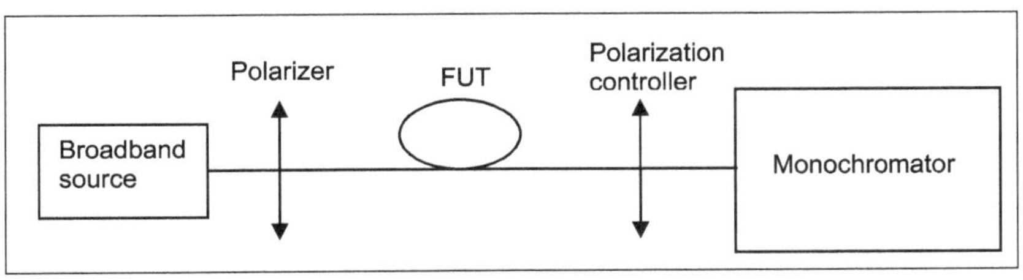

Fixed analyzer setup.

The fixed analyzer method, also called the wavelength scanning method, involves launching monochromatic polarized light into the fiber under test and measuring the spectral transmission with another polarizer. The linear birefringence of a short fiber induces a sinusoidal transmission response when scanning the wavelength.

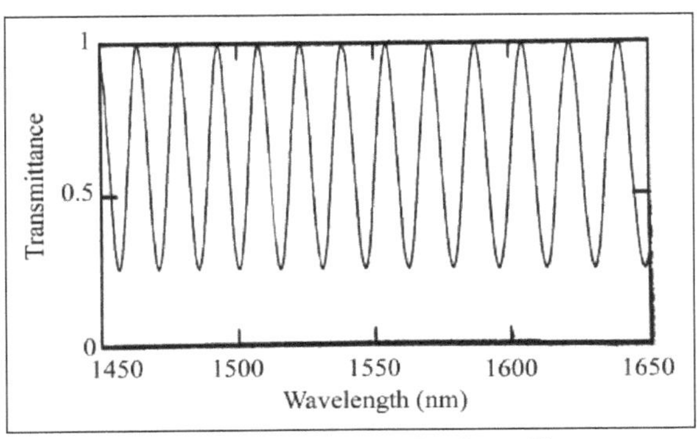

Wavelength scanning of a short fiber.

The calculation of PMD on a short fiber is straightforward, it is the inverse of the fringe spacing (in frequency) in the wavelength scanning method. But fiber lengths in networks are much longer than the coupling length. These changes the aspect of the signal obtained, as illustrated in the example below showing a 50 km fiber with a 0.1 ps/km specific PMD and a 1 km coupling length.

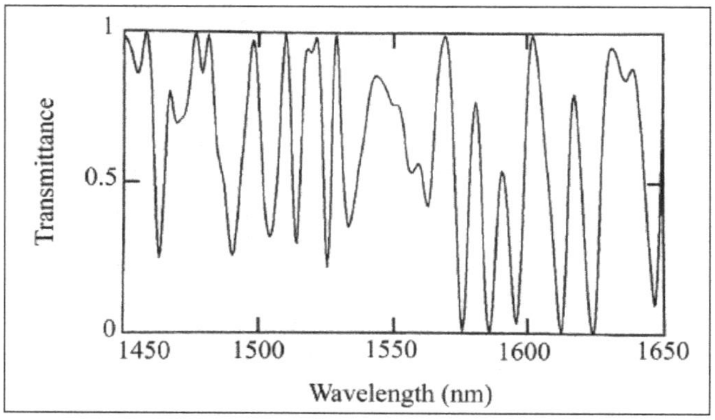

Wavelength scanning of a long fiber.

The most striking feature compared to a short fiber is that the spectral transmission is no longer sinusoidal. Around some wavelengths (1530 nm in our example), the fringe separation is narrower, indicating that the effect of PMD is higher around these wavelengths. The definition of PMD is not as straightforward as it was for short fiber. One method uses fringe counting over the wavelength span. A PMD value can be established this way, although it is not always clear what should be considered as a fringe. For instance, should the small peak on the graph around 1555 nm be counted or not? Another method uses the Fourier transform of the transmission spectrum and is technically very similar to interferometry.

The fixed analyzer method is limited to a 60 ps delay range and is very sensitive to the fiber movement, thus causing uncertainty. Averaging is required for better accuracy, but this significantly increases the measurement time. Also, as this method requires step-by-step wavelength scanning, it is necessary to ensure that the wavelength step is small enough with regard to the PMD to be measured. The Jones matrix method uses a tunable laser source, a polarization controller, and a polarization analyzer. At each wavelength, the polarization controller is scanned and a mathematical calculation is performed using the Jones matrix formalism (hence the name of the method). From this, the Differential Group Delay - the maximum delay that can be encountered over polarization at that wavelength is derived.

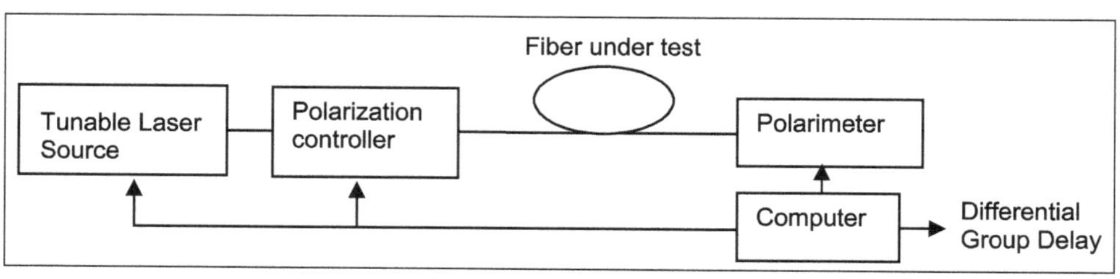

Jones matrix method setup.

For a short fiber (shorter than the coupling length), Differential Group Delay is independent of wavelength. For long fibers, DGD is no longer independent of wavelength, as illustrated in the example below of a 50 km fiber with a 0.1 ps/ km specific PMD and a 1 km coupling length. The PMD definition must take this wavelength dependence into account. PMD is computed as the mean value for DGD over a specified wavelength range, typically ±40 nm around an operating wavelength.

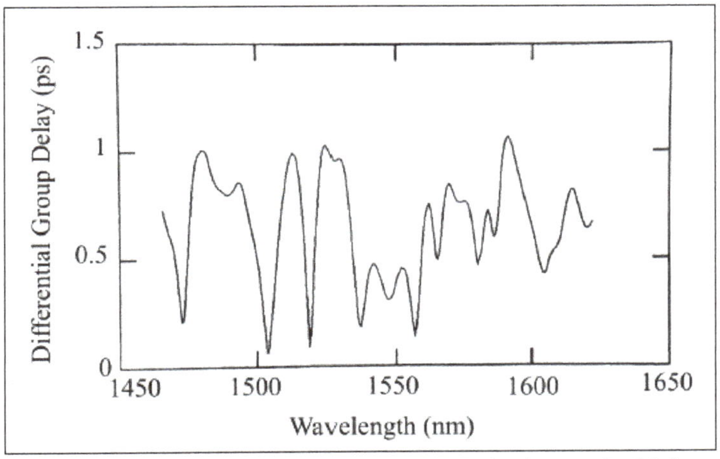

Differential Group Delay versus wavelength (Jones matrix eigenanalysis) for a long fiber.

The Jones matrix method gives an indication of the Differential Group Delay dependence with wavelength. Using a narrow band tunable laser, the minimum measurable PMD value is not limited. As in the fixed analyzer method, the Jones matrix eigenanalysis requires that the fibers not be moved during the measurement procedure, since this would create measurement artifacts. It is therefore more suitable for laboratory applications. This method also requires step-by-step wavelength scanning. It is necessary to ensure that the wavelength step is small enough with regard to the PMD to be measured. The Jones matrix eigenanalysis is a complex method that requires polarization scanning. The time required for testing is usually long: about 10 minutes for each measurement. Finally, it is an expensive solution, limited to the C&L bands, due to the need to use a tunable laser source.

The Interferometric Method uses a polarized broadband source and a Michelson scanning interferometer. The interferometer can be seen as a Fourier transform spectrum analyzer, or more simply as a correlator for the PMD-delayed wave trains.

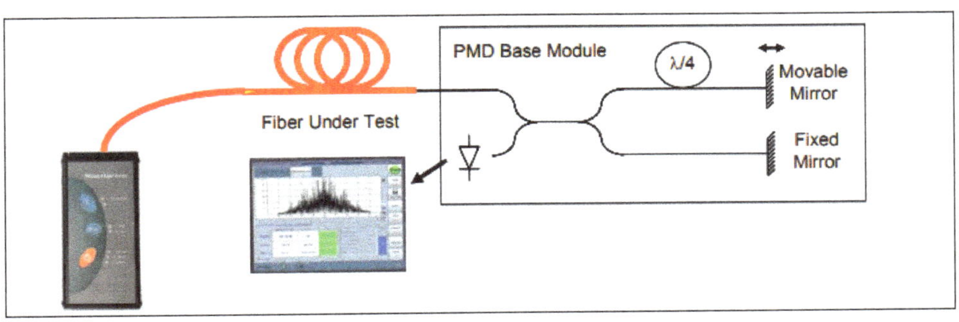

Interferometric method setup.

On a short fiber, the interferogram envelope shows a single central peak (which corresponds to the autocorrelation peak of the source) and two satellite peaks, as represented in the figure below. The interferometer unbalance between the central peak and the satellite peaks results in the PMD. The interferogram can also be seen as the Fourier transform of the transmitted spectrum of the wavelength scanning method, taken over the spectral range of the source.

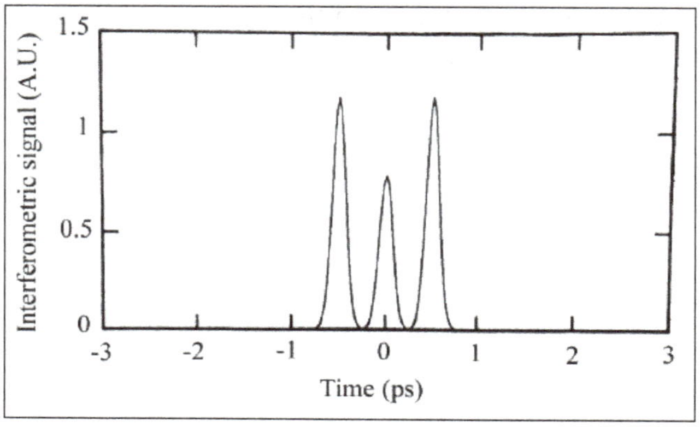

Interferogram envelope for a short fiber.

For longer randomly mode-coupled fibers, the interferogram is close to a Gaussian function multiplied by a noise after removing the auto-correlation peak. From a mathematical point of view, the signal is the envelope of the Fourier transform of the transmission spectrum resulting from the wavelength scanning method. PMD is given by the width of the envelope.

The auto-correlation peak does not contain any information related to PMD measurement. It can be removed either by computation, or by adding a quarter-wave-plate inside the interferometer to create destructive interferences when both arms of the interferometer have the same length. This second method enables a higher dynamic range, resulting in the ability to test longer links, a higher accuracy, and better repeatability.

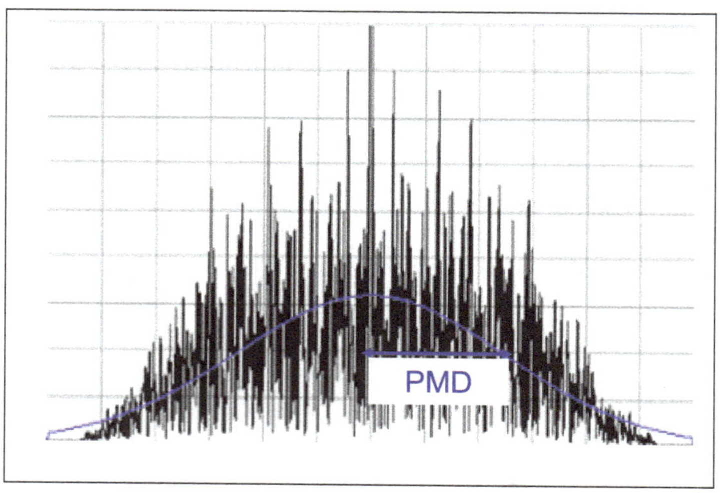

Interferogram envelope for a long fiber.

This method is well suited for field measurement requirements. The test instruments are rugged and portable, and can operate on batteries. The interferometric method is insensitive to fiber moves. It provides accurate measurements in a few seconds, and does not require averaging. High power broadband sources can be used to increase the dynamic range, allowing measurement of long links up to 250 km. This method also allows PMD measurements through multiple amplifiers such as EDFAs. The minimum measurable PMD value is limited in this method, depending on the spectral width of the broadband source. However, sources with FWHM >50 nm will allow PMD measurement down to 0.05 ps, which is quite acceptable in the field.

The interferometric method has been standardized in two versions: The TINTY (traditional interferometry) version, as described above, and the GINTY (generalized interferometry) version. GINTY differs from TINTY by adding polarization scramblers and an analyzer at the input and the output of the fiber under test. This is another way to remove the auto-correlation peak from the interferogram and then measure long links having EDFAs. However, because scramblers are used, averaging a large number of measurements is necessary to achieve the full benefit of this method. As was explained above, technical solutions also exist in the TINTY method to remove this central peak using methods other than computation.

Table: Comparison table of PMD measurement methods.

	Fixed analyzer method	Jones matrix method	Interferometric methods	
			TINTY	GINTY
Application	Any link in the field (measurement through EDFAs possible)	Laboratory	Any link in the field (measurement through EDFAs possible)	Any link in the field (measurement through EDFAs possible)
Measurement constraint	High sensitivity to fiber movement	High sensitivity to fiber movement C&L bands only		
Measurement range	0.08 ps to 60 ps	0.03 ps to 50 ps	0.04 ps to 80 ps	0.04 ps to 115 ps
Dynamic range	< 50 dB	< 50 dB	> 60 dB (high dynamic both at 1310 nm and 1550 nm)	< 50 dB
Measurement speed	Fast for 1 measurement but averaging is required	Slow	Fast	Fast for 1 measurement but averaging is required
Cost	Low	High	Low	Low

Polarization

Understanding and manipulating the polarization of light is crucial for many optical applications. Optical design frequently focuses on the wavelength and intensity of light, while neglecting its polarization. Polarization, however, is an important property of light that affects even those optical systems that do not explicitly measure it. The polarization of light affects the focus of laser beams, influences the cut-off wavelengths of filters, and can be important to prevent unwanted back reflections. It is essential for many metrology applications such as stress analysis in glass or plastic, pharmaceutical ingredient analysis, and biological microscopy. Different polarizations of light can also be absorbed to different degrees by materials, an essential property for LCD screens, 3D movies, and your glare-reducing sunglasses.

Light is an electromagnetic wave, and the electric field of this wave oscillates perpendicularly to the direction of propagation. Light is called unpolarized if the direction of this electric field fluctuates randomly in time. Many common light sources such as sunlight, halogen lighting, LED spotlights, and incandescent bulbs produce unpolarized light. If the direction of the electric field of light is well defined, it is called polarized light. The most common source of polarized light is a laser.

Depending on how the electric field is oriented, we classify polarized light into three types of polarizations:

- Linear polarization: The electric field of light is confined to a single plane along the direction of propagation.
- Circular polarization: The electric field of light consists of two linear components that are perpendicular to each other, equal in amplitude, but have a phase difference of $\pi/2$. The resulting electric field rotates in a circle around the direction of propagation and, depending on the rotation direction, is called left- or right-hand circularly polarized light.
- Elliptical polarization: The electric field of light describes an ellipse. This results from the combination of two linear components with differing amplitudes and/or a phase difference that is not $\pi/2$. This is the most general description of polarized light, and circular and linear polarized light can be viewed as special cases of elliptically polarized light.

The two orthogonal linear polarization states that are most important for reflection and transmission are referred to as p- and s-polarization. P-polarized (from the German parallel) light has an electric field polarized parallel to the plane of incidence, while s-polarized (from the German senkrecht) light is perpendicular to this plane.

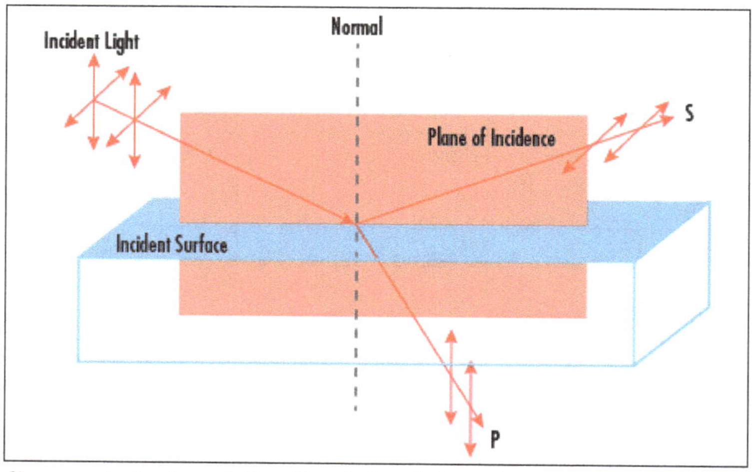

P and S are linear polarizations defined by their relative orientation to the plane of incidence.

Manipulating Polarization

Polarizers

In order to select a specific polarization of light, polarizers are used. Polarizers can be broadly divided into reflective, dichroic, and birefringent polarizers.

Reflective polarizers transmit the desired polarization while reflecting the rest. Wire grid polarizers are a common example of this, consisting of many thin wires arranged parallel to each other. Light that is polarized along these wires is reflected, while light that is polarized perpendicular to these wires is transmitted. Other reflective polarizers use Brewster's angle. Brewster's angle is a specific angle of incidence under which only s-polarized light is reflected. The reflected beam is s-polarized and the transmitted beam becomes partially p-polarized.

Dichroic polarizers absorb a specific polarization of light, transmitting the rest; modern nanoparticle polarizers are dichroic polarizers. Birefringent polarizers rely on the dependence of the refractive index on the polarization of light. Different polarizations will refract at different angles and this can be used to select certain polarizations of light.

Unpolarized light can be considered a rapidly varying random combination of p- and s-polarized light. An ideal linear polarizer will only transmit one of the two linear polarizations, reducing the initial unpolarized intensity I_o by half,

$$I = \frac{I_o}{2}$$

For linearly polarized light with intensity I_o, the intensity transmitted through an ideal polarizer, I, can be described by Malus' law,

$$I = I_o \cos^2 \theta$$

where θ is the angle between the incident linear polarization and the polarization axis. We see that for parallel axes, 100% transmission is achieved, while for 90° axes, also known as crossed polarizers, there is 0% transmission. In real world applications the transmission never reaches exactly 0%, therefore, polarizers are characterized by an extinction ratio, which can be used to determine the actual transmission through two crossed polarizers.

Waveplates

While polarizers select certain polarizations of light, discarding the other polarizations, ideal waveplates modify existing polarizations without attenuating, deviating, or displacing the beam. They do this by retarding (or delaying) one component of polarization with respect to its orthogonal component. To help you determine which waveplate is best for your application. Correctly chosen waveplates can convert any polarization state into a new polarization state, and are most often used to rotate linear polarization, to convert linearly polarized light to circularly polarized light or vice versa.

Polarization Vectors

Maxwell's equations admit plane wave solutions,

$$\vec{E} = \vec{E}_o e^{i(\vec{k}\cdot\vec{x}-\omega t)}, \quad \vec{B} = \vec{B}_o e^{i(\vec{k}\cdot\vec{x}-\omega t)}$$

Here, $\vec{E}_o$ and $\vec{B}_o$ are called the polarization vectors for the electric and magnetic fields. These are complex 3 dimensional vectors. The wavevector $\vec{k}$ and angular frequency ω are real and in the vacuum are related by $\omega = c|\vec{k}|$. This relation implies that electromagnetic waves are dispersion less with velocity c: the speed of light. In materials, like a prism, light can have dispersion. In addition, we found that for plane waves,

$$\vec{B}_o = \frac{1}{\omega}\left(\vec{k} \times \vec{E}_o\right)$$

This equation implies that the magnetic field in a plane wave is completely determined by the electric field. In particular, it implies that their magnitudes are related,

$$|\vec{E}_o| = c|\vec{B}_o|$$

and that,

$$\vec{k}\cdot\vec{E}_o = 0,\ \ \vec{k}\cdot\vec{B}_o = 0,\ \ \vec{E}_o\cdot\vec{B}_o = 0$$

In other words, the polarization vector of the electric field, the polarization vector of the magnetic field, and the direction $\vec{k}$ that the plane wave is propagating are all orthogonal. To see how much freedom there is left in the plane wave, it's helpful to choose coordinates. We can always define the $\hat{z}$ direction as where $\vec{k}$ points. When we put a hat on a vector, it means the unit vector pointing in that direction, that is $\hat{z} = (0,0,1)$. Thus the electric field has the form,

$$\vec{E} = \vec{E}_o e^{i\omega\left(\frac{z}{c} - t\right)}$$

which moves in the z direction at the speed of light. Since $\vec{E}_o$ is orthogonal to $\vec{k}$ we can write also write,

$$\vec{E}_o = (E_x, E_y, 0)$$

with E_x and E_y complex numbers.

The two complex amplitudes E_x and E_y each have magnitudes and phases. We can write these explicitly as $E_x = |E_x|e^{i\phi_x}$ and $E_y = |E_y|e^{i\phi_y}$. So one needs four real numbers to specify the polarization vector. We usually separte the phases into an overall phase and the difference in phase between E_x and E_y, $\varphi = \varphi_x - \varphi_y$. The reason for doing this is that as t and z change the phases of E_x and E_y change in the same way while φ is unaffected. We don't usually care about this overall phase, or about the overall magnitude $E = |E_x|^2 + |E_y|^2$. Thus, to specify a polarization vector, we talk about the relative size of E_x and E_y and the phase difference $\varphi = \varphi_x - \varphi_y$.

Don't let the complex polarization vectors confuse you. We can just as,

$$\vec{E} = E_x\hat{x}\cos(kz - \omega t + \phi_x) + E_y\hat{y}\cos(kz - \omega t + \phi_y)$$

where you see the four real numbers specifying the polarization explicitly. Exponentials just make a lot of the algebra easier.

Linear Polarization

We say a plane wave is linearly polarized if there is no phase difference between E_x and E_y. We can write linear polarizations as,

$$\vec{E}_o = (E_x, E_y, 0)$$

and choose the overall phase so that E_x and E_y are real numbers. If $E_y = 0$ but $E_x \neq 0$, we have,

$$\vec{E} = E_o\hat{x}e^{i(kz-\omega t)}$$

with $E_o = |\vec{E}_o|$ just a number now. Then, from Equation $\vec{B}_o = \frac{1}{\omega}(\vec{k} \times \vec{E}_o)$ since $\hat{z} \times \hat{x} = \hat{y}$,

$$\vec{B} = \frac{1}{c} E_o \hat{y} e^{i(kz-\omega t)}$$

This configuration is said to be linear polarized in the x direction.

Similarly we can have,

$$\vec{E} = E_o \hat{y} e^{i(kz-\omega t)}$$

and using $\hat{z} \times \hat{y} = -\hat{x}$:

$$\vec{B} = -\frac{1}{c} E_o \hat{x} e^{i(kz-\omega t)}$$

This configuration is said to be linear polarized in the y direction. Note that in both cases, the magnetic field is given by rotating the electric field 90° counterclockwise in the x-y plane and dividing by c.

More generally, for any unit vector $\hat{v}$ we can have,

$$\vec{E} = E_o \hat{v} e^{i(\vec{k}\cdot\vec{x}-\omega t)}$$

which is linearly polarized in the $\hat{v}$ direction. The magnetic field will be polarized in direction 90° behind the electric field.

Remember, we always implicitly want to have the real part of these fields. So, for linearly polarized light in the x direction, the fields are actually,

$$\text{Re}\left[\vec{E}\right] = \left(E_o \cos(kz-\omega t), 0, 0\right), \ \text{Re}\left[\vec{B}\right] = \left(0, \frac{E_o}{c}\cos(kz-\omega t), 0\right)$$

Note that there is no x or y dependence in these solutions: the fields are completely uniform in x and y – they are plane waves. At each point on the plane, the electric field points in the $\hat{x}$ direction with the same magnitude. This magnitude varies as we move in z and in t but is always uniform in the plane.

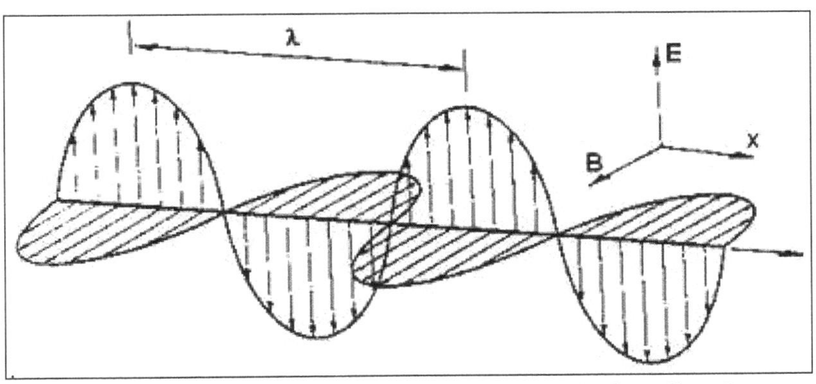

Fields vary as they move along for a wave moving in the x direction.

Circular Polarization

What if the components of the electric field are not in phase? First suppose they have the same magnitude but are a quarter wavelength out of phase, so $\phi_x - \phi_y = \frac{\pi}{2}$. Then,

$$\vec{E}_o = \left(E_o, E_o e^{i\frac{\pi}{2}}, o \right) = (E_o, iE_o, o)$$

Thus, $\vec{E} = \left(E_o e^{i(kz-\omega t)}, iE_o e^{i(kz-\omega t)}, o \right)$

Taking the real part gives the actual electric field,

$$\text{Re}\left[\vec{E}\right] = \left(E_o \cos(kz-\omega t),\ -E_o \sin(kz-\omega t),\ o\right)$$

This is called left-handed circularly polarized light.

What does it look like? Well, at t = z = o, the field is $\text{Re}(\vec{E}) = (E_o, o, o)$ pointing in the x direction. A little farther along, when $kz = \frac{\pi}{2}$, still at t = o, then $\text{Re}(\vec{E}) = (o, -E_o, o)$ which points in the negative $\hat{y}$ direction. Farther along still, when kz = π, it points along $-\hat{x}$, and then $\hat{y}$ and so on. Finally, after a full wavelength, it goes back to $\hat{x}$. Thus, as we move along z, the polarization rotates clockwise in the x-y plane. Equivalently, at a given z, as time progresses it also rotates in the x-y plane.

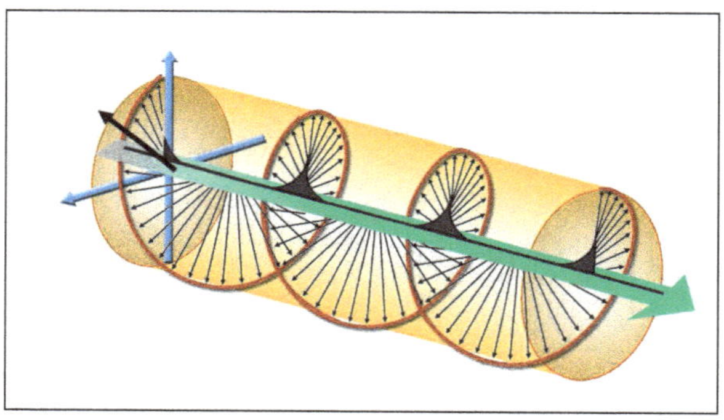

Circularly polarized light changes the direction of its polarization as it moves.

Similarly, taking $\phi_x - \phi_y = -\frac{\pi}{2}$ gives $\vec{E}_o = (E_o, -iE_o, o)$

$$\text{Re}\left[\vec{E}\right] = \left(E_o \cos(kz-\omega t), E_o \sin(kz-\omega t),\ o\right)$$

which is right-handed circularly polarized light. In this case, at t = o and kz = o, the polarization points in the $\hat{x}$ direction. A quarter wavelength farther, it points in the $\hat{y}$ direction, and so one. So this field rotates counterclockwise in the x-y plane. Note that when we add left and right handed polarizations we,

$$\vec{E}_o = (E_o, iE_o, o) + (E_o, -iE_o, o) = (2E_o, o, o)$$

which is linearly polarized in the $\hat{x}$ direction. Similarly, subtracting them gives linear polarization in the $\hat{y}$ direction. Thus circular and linear polarizations are not linearly independent. Indeed, any possible polarization can be written as a linearly combination of left-handed and right-handed circularly polarized light.

Circularly polarized light has angular momentum $\pm e_0 |\vec{E}|^2 \hat{k}$, with the positive sign for left-handed and the negative sign for right-handed. Linearly polarized light carries no angular momentum. This is easy to understand because it is a sum of left and right handed light. As an analogy, suppose you have two tops, one spinning clockwise and one spinning counterclockwise; the two-top system has no net angular moment.

Light cannot have arbitrarily small intensity. The smallest intensity light can have is a single photon. Thus the photon itself must be polarized. A single circularly polarized photon has angular momentum $\vec{J} = \pm \frac{h}{2\pi} \hat{k}$. It may seem surprising that photons which are point-like particles with no substructure can have angular momentum on their own. It is a very odd fact, but true nonetheless. Photons have spin. We say photons are particles of spin 1. The two signs for the angular momentum correspond to the two helicities of the photon. To complete the discussion of polarization, we can consider also varying the magnitudes of the different components and the phase. The most general parameterization is,

$$\vec{E}_0 = E_x \hat{x} \cos(kz - \omega t) + E_y \hat{y} \cos(kz - \omega t + \phi)$$

When E_x and E_y are different and the relative phase φ is nonzero, the polarization changes in magnitude as it rotates in the x-y plane. It thereby describes an ellipse, and this is called elliptical polarization. Linear polarization corresponds to $\varphi = 0$. Circular polarization corresponds to $\phi = \pm \frac{\pi}{2}$ and $E_x = E_y$.

Polaroid Film

One way to polarize light is using a Polaroid film. The first such film was invented by Edwin Land in 1928, while an undergraduate at Harvard. He came up with a way to align polymer molecules in a thin sheet into long needlelike strands. When an electric field acts on the electrons in those strands, it can only move the electrons up and down in the strand direction, but not perpendicular to the strands. Thus the only polarization which can pass through is linearly polarized in the direction perpendicular to the strips.

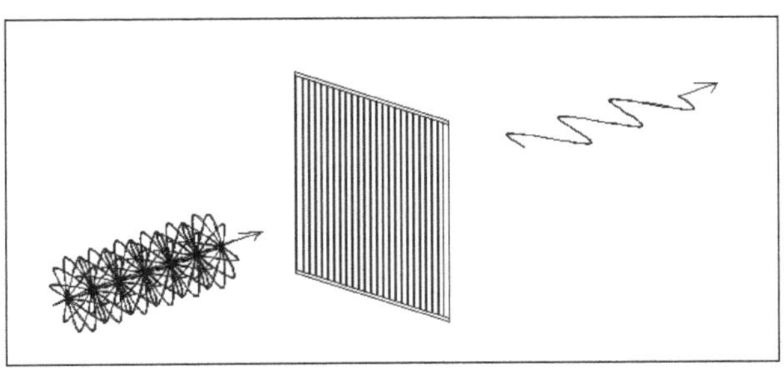

Polarizing film absorbs the electric field in the direction of the strips.

Land called his invention Polaroid film, patented it, and then founded the Polaroid Corporation here in Cambridge. One of the primary applications at the time was sunglasses. His next most famous invention was the Land instant camera, now known as the Polaroid camera. The Polaroid camera has nothing to do with polarization. Back to polarizing film. Say the film has the strips in the y direction and a plane wave comes in along the z direction with an arbitrary polarization,

$$\vec{E}_{init} = E_x \hat{x} e^{i(kz-\omega t)} + E_y \hat{y} e^{i(kz-\omega t+\phi)}$$

Since the polarizer has strips in the y direction, it absorbs the $\hat{y}$ components of the electric field. Thus, what exits the polarizer must be,

$$\vec{E}_{final} = E_x \hat{x} e^{i(kz-\omega t)}$$

Thus the polarizer takes any initial polarization and turns in into linear polarization in the x direction.

For example, say the field is linearly polarized at an angle θ to the direction where the field would just go through. That is,

$$\vec{E}_{init} = E_o \left(\cos\theta \hat{x} + \sin\theta \hat{y}\right) e^{i(kz-\omega t)}$$

Then if we put it through the polarizer with strips in the y direction we would find,

$$E_{init} = E_o \cos\theta \hat{x} e^{i(kz-\omega t)}$$

So that the result is still linearly polarized but now in the $\hat{x}$ direction with magnitude $E_o\cos\theta$. Since the intensity of radiation is proportional to the square of the field, we would find that a polarizer oriented at an angle θ to the direction of propagation reduces the intensity,

$$I_{final} = I_{initial} \cos^2\theta$$

This is known as Malus' law.

Polarization in the Real World

The key facts which let us understand why natural light is polarized are that the electric field makes charged particles move in the direction of $\vec{E}$ and moving charged particles produce electric fields along their direction of motion. Fact and the consequent absorption of energy of the electric field is the mechanism behind the Polaroid film. During the day, sunlight is constantly being absorbed and emitted by molecules (mostly water) in the atmosphere. Let's say z is the vertical direction (towards the sky), so the sky above us is the x-y plane. Let's say the sun is in the x direction. Thus plane waves from the sun have $\vec{k}$ in the $\hat{x}$ direction and are polarized in the y-z plane.

Thus, the sky molecules can only get pushed in the y and z directions from sunlight. Molecules moving in the z direction emit light in the x-y plane, that is, off into the sky, but not to us. Molecules excited in the y direction however can emit light going directly off in z down to us. Since all this light comes from motion in the y-z plane, the light will all be linearly polarized in this plane. Thus the sky directly above us, at right angles to the sun, is polarized. Similar arguments show that light from some other angle will have some polarization, but not be completely polarized. Looking through the sky directly at the sun, the light should not be polarized at all.

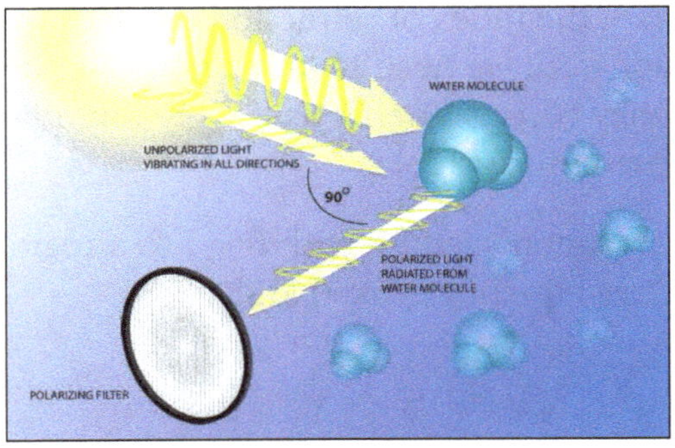

Polarization of the sky.

Another important example is light from a reflection. Here the story is nearly identical with the maximum polarization coming if the source is at a right angle to the viewer. In this case, the index of refraction of the two materials involved (air and lake for example) play a role and the angle is not exactly 90 degrees. Instead it is called Brewster's angle.

Polarization from reflection.

So reflected light is polarized. Most light around us is not polarized. So for example, if you are on a boat and light is coming in from the sky (not just the sky directly above you), and also from the water, only the light reflected off the water will be polarized. Thus if you put on polarizing sunglasses, you can filter out the reflected light, and see more clearly the things around you. Similarly, reflected light coming off the ground will be polarized horizontally. Polarizing sunglasses generally try to filter out horizontally polarized light.

Index of Refraction

The speed of light c is the speed of light in the vacuum. In air, light goes slightly slower than c. More generally, light travels at a speed,

$$v = \frac{1}{n} c$$

with n called the index of refraction. The physical origin of the index of refraction is interesting. Light comes in and excites molecules. These molecules then radiate more light which interferes with the incoming light. The phases of the incoming and radiated light align in such a way as to make the light appear to be slower. For now, let's accept the index of refraction as a phenomenological fact.

When light hits a boundary where the index of refraction changes, we have to solve the wave equation with the boundary conditions. All we need to know for now is that the incident and transmitted wave have the same frequency (this is always true – the boundary knows about the frequency but not the wavelength of the incoming wave).

Say light with frequency ω goes from a vacuum with n = 1 to a region with n = 2 and then back to the vacuum with n = 1. Since the frequency is fixed and,

$$\omega = vk = \frac{c}{n}\frac{2\pi}{\lambda}$$

we must have,

$$n_1\lambda_1 = n_2\lambda_2$$

So that as n goes up (at fixed frequency), λ must go down.

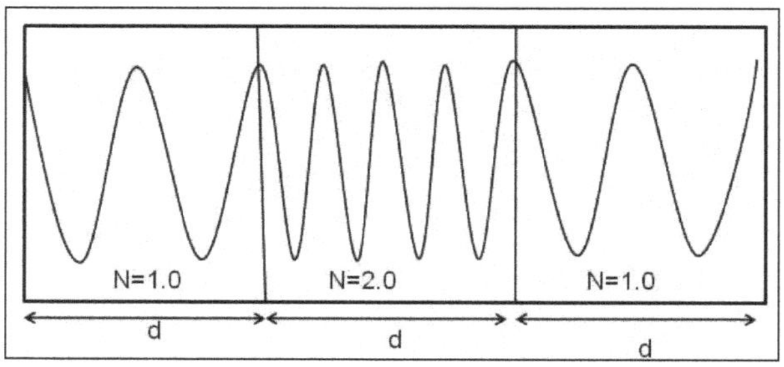

Plane wave entering and emerging from a medium with different index of refraction. Figure illustrates how light going slower means it goes through more oscillations over the same distance.

Many materials are birefringent, meaning they have different indices of refraction in different directions. A linear polarizer is an extreme example: it has n = ∞ in one direction, so that light simply does not propagate for one polarization. The mineral calcite is a naturally occurring birefringent material. Calcite's two indices of refraction are,

$$n_1 = 1.66, \quad n_2 = 1.49$$

These indices of refraction are determined by the atomic structure of the mineral and associated with the primary axes of the crystal structure. You can see the birefringence by looking at some writing through a chunk of calcite. You will see two images. The two images are due to the different polarizations of light refracting differently in the crystal. As you rotate the crystal, you can see how one polarization is refracting more than the other.

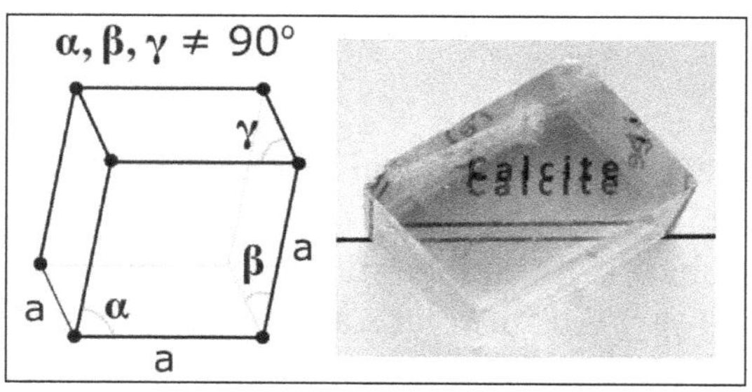

Calcite is naturally birefringent. When looking through a calcite crystal you see two images because the two polarizations of light refract differently.

Quarter and Half-wave Plates

Suppose we have a plane wave moving in the $\hat{z}$ direction whose electric field is polarized in the $\hat{x}+\hat{y}$ direction. Say this wave passes through a calcite crystal of thickness L whose optical axes are aligned with the $\hat{x}$ and $\hat{y}$ directions. The initial electric field is,

$$\vec{E}_{init}(z,t) = E_o(\hat{x}+\hat{y})e^{i\omega\left(t-\frac{n}{c}z\right)}$$

where n is the index of refraction in air. Inside the crystal, the two components propagate with different speeds, so (assuming zero reflection),

$$\vec{E}_{inside}(z,t) = E_o\hat{x}e^{i\omega\left(t-\frac{n_x}{c}z\right)} + E_o\hat{y}e^{i\omega\left(t-\frac{n_y}{c}z\right)}$$

where n_x and n_y are the two indices of refraction of the calcite.

In particular, at z = L we have,

$$\vec{E}_{inside}(L,t) = E_o\hat{x}e^{i\omega\left(t-\frac{n_x}{c}L\right)} + E_o\hat{y}e^{i\omega\left(t-\frac{n_y}{c}L\right)}$$

Thus there is a net phase difference of,

$$\Delta\phi = \frac{\omega}{c}L(n_x - n_y)$$

between the x and y components of a field. Then when the light exits and both indices of refraction are n again, the outgoing wave has electric field,

$$\vec{E}_{final} = E_o(\hat{x}+\hat{y}e^{i\Delta\phi})e^{i\omega\left(t-\frac{n}{c}z\right)}$$

From this calculation, we can now deduce how to use calcite to turn linearly polarized light into circularly polarized light. To do so, we must choose the length L so that $\Delta\phi = \frac{\pi}{2}$. This is easy to do if we know n_x and n_y using equation $\Delta\phi = \frac{\omega}{c}L(n_x - n_y)$. A material which rotates the phase by $\frac{\pi}{2}$ which is a quarter of a circle, is known as a quarter-wave plate.

A few things to note about quarter wave plates or more generally use birefringent materials to polarize light:

- The phase $\Delta\varphi$ depends linearly on the frequency ω. So you need different lengths for different frequencies. In most materials, n_x and n_y also depend on ω.
- Light which is linearly polarized along one of the axes of the crystal will remain linearly polarized. That is, the orientation of the quarter wave plate is important.

In practice, if one has a coherent, monochromatic, polarized source, such as a laser, one knows the frequency and the direction of polarization. Then one can rotate between linear and circular polarization using wave plates.

If one places two quarter-wave plates together, or equivalently takes a crystal or length 2L, one gets a half-wave plate. The first L of it adds a phase $e^{i\frac{\pi}{2}} = i$, rotating linear polarization to circular polarization; the second L adds another phase $e^{i\frac{\pi}{2}} = i$, rotating circular back to linear. However, the net phase changes is $e^{i\pi} = -1$ so $(E_x, E_y) \rightarrow (E_{x,} -E_y)$ and the final polarization is now rotated 90° from the original polarization. Thus half-wave plates can rotate the polarization of linearly polarized light. A crystal of length 4L will have $\Delta\varphi = 2\pi$ and return the polarization to its initial value. This is called a full-wave plate.

One application of the use of circularly polarized light is in 3D movies. Ever wonder what those glasses they give you at IMAX do? The IMAX Theater projects two movies at once, one with left-hand circular polarization and the other with right-handed circular polarization. Your 3D glasses have one lens which can see transmit the left-handed light and the other lens transmits only right-handed. Thus you see different images with each eye which gives the illusion of depth. Why do you think they don't use linearly polarized light?

For another application, consider that the index of refraction along axes in some materials can be controlled by applying electric and magnetic fields. In the case of electric fields, this effect is called electro-optic effect. Then the index of refraction is $n(\vec{E})$. In the case of magnetic fields, this effect is called the Faraday effect, and the index of refraction is $n(\vec{B})$. These effects are exploited in electro-optic devices which allow fast and precise control of the polarization of light.

An example of an electro-optic device is the rapatronic camera. In an ordinary camera, where the shutter is mechanical, the shutter speed is limited to about 1 millisecond. In a rapatronic camer, the shutter is composed of two crossed polarizers with a birefringent material exhibiting the electro-optic effect in between the polarizers. With the voltage on, the material acts as a half-wave plate so that all the light is transmitted through the shutter. If the voltage is switched off, which can be done on the order of microseconds or faster, the material has no effect on the polarization. This causes the shutter to be closed. Rapatronic cameras were heavily developed during the 1940's to take pictures of atomic bomb detonations very shortly (10^{-6} s) after the detonation. This allowed scientists to learn about how well the bomb, in particular its detonation, was working.

References

- Interference-of-Waves: physicsclassroom.com, Retrieved 17, May 2020
- Single-slit-diffraction: byjus.com, Retrieved 03, August 2020
- Disp-in-Opt-Fibers-PMD-CD, White-Paper, Technical-Notes: cdn-anritsu.com, Retrieved 25, June 2020
- Introduction-to-polarization, knowledge-center: edmundoptics.com, Retrieved 19, March 2020
- Polarization: harvard.edu, Retrieved 21, July 2020

3

Geometrical Optics: A Comprehensive Study

Geometrical optics is the branch of optics that studies the phenomena where light behaves as rays. Various phenomena such as reflection, specular reflection, diffuse reflection, signal reflection, total internal reflection, reflection phase change, refraction and refractive index. This chapter has been carefully written to provide an easy understanding of the varied facets of geometrical optics.

Geometrical optics deals with the propagation of light in a straight line and phenomena such as reflection, refraction, polarization, etc. A ray of light gives the direction of propagation of light. In the absence of an obstacle, the rays advance in a straight line without changing direction. When light meets a surface separating two transparent media, reflection and refraction occur and the light rays bend.

Every real system experiences diffraction effects, so geometric optics is necessarily an approximation. However, the simplicity arising from treating only rays which move in straight lines affords many uses.

Reflection at Smooth Surfaces

A light ray is reflected by a smooth surface in accordance with the 2 laws of reflection:

1. The angle of incidence is always equal to the angle of reflection.
2. The incident ray, the reflected ray, and the normal to the reflecting surface are coplanar. The images formed by a plane mirror are equidistant from the distance between the object and the mirror.

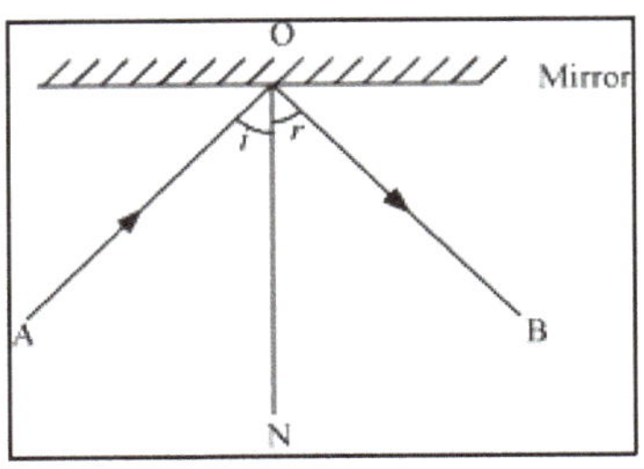

Example

An object is placed in front of the mirror. If the mirror is moved away from the object through a distance of x by how much distance will the image move?

The answer will be added shortly.

Our final answer is $2x$.

A ray of light falls on a plane mirror. Show that if the mirror is tilted through an angle of θ, then the reflected ray tilts through an angle of 2θ.

The laws of reflection are the same for plane and curved surfaces. A normal can be drawn from any point of the curved surface by first drawing the tangent plane from that point and then drawing the line perpendicular to that plane. Angles of incidence and reflection are defined from this normal. The angle of incidence is equal to angle of reflection.

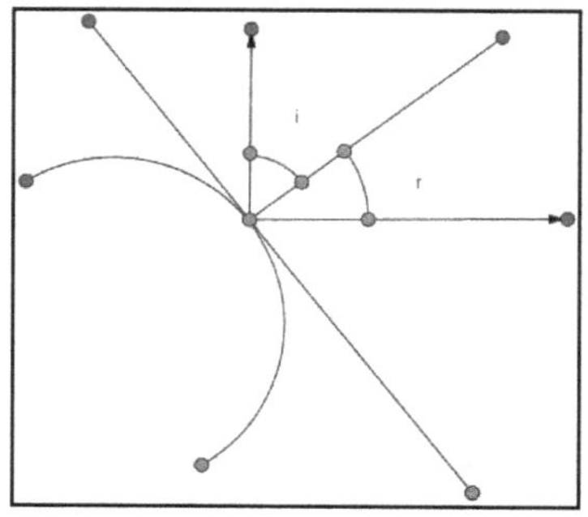

Spherical Mirrors

A spherical mirror is a part cut from a hollow sphere. The spherical mirrors are generally constructed from glass. One surface of the glass is silvered. If reflection takes place at the convex surface, it is called a convex mirror and if it takes place on the concave surface, it is called a concave mirror.

Terms used in spherical mirrors:

- Center of curvature (C): it is the center of the sphere from which the mirror is made.
- Focus (f): it is the point at which the rays meet (either virtually or physically). Although we have not verified, from further exploration we will understand that $R = 2f$.
- Pole (P): it is the surface of the mirror on which light falls.
- Principle axis (PA): it is the imaginary line which is perpendicular to the pole.
- Paraxial rays: These are rays parallel and close to the principal axis.
- u : This is the distance between the object and the pole and is sometimes referred to as the object distance.
- v : This is the distance between the pole and the image and is sometimes referred to as the image distance.

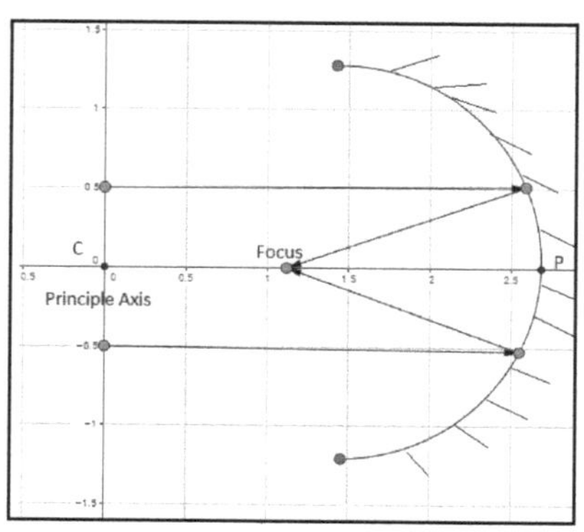

Here $(0, 0)$ is the C and the principal axis is the x-axis.

Sign Conventions

Cartesian Sign Conventions

- The pole is taken to be the origin.
- The direction in which the light moves is taken as the positive direction.
- The height above the principal axis is taken as positive while the height below it is taken as negative.

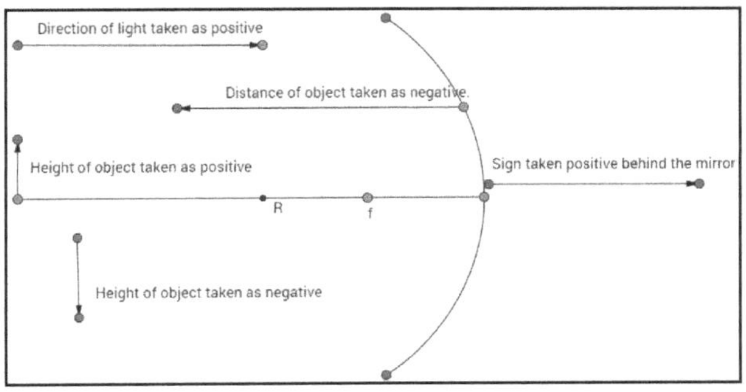

Mirror Formula

Theorem,

The main formula used for solving problems on mirrors is,

$$\frac{1}{v}+\frac{1}{u}=\frac{1}{f}.$$

While solving problems, do not forget to apply sign conventions and make sure you always take the object distance, u as *negative only*. the examples below which will make you comfortable with solving the problems based on the mirror formula.

Example

If the distance between the object and the mirror is $10\,\text{cm}$ and its focal length is $5\,\text{cm}$ given it is a convex mirror, find the distance between the mirror and the object.

We are given $u=-10\,\text{cm}$ and $f=5\,\text{cm}$

Applying the mirror formula, we get

$$\frac{1}{v}+\frac{1}{u}=\frac{1}{f}$$

$$\frac{1}{v}+\frac{1}{-10}=\frac{1}{5}$$

$$\frac{1}{v}=\frac{3}{10}$$

$$\Rightarrow v=3.33\,\text{cm}.$$

THEOREM

The magnification of a mirror is given by $\frac{h_i}{h_0}$, where h_i is height of the image and h_0 is height of the object.

Magnification is also given by $-\frac{v}{u}$.

Again we should remember not to forget applying the sign conventions.

If magnification is $-ve$ it means the formed image is inverted.

Example

If the image formed by a spherical mirror is half the size of the object and it is virtual, then find the distance between the mirror and the image given that the object's distance is 16cm.

We are given $m = \frac{h_i}{h_0} = \frac{1}{2} = \frac{-v}{u}$.

So, $u = -2v = -16 \Rightarrow v = 8\text{cm}$.

Refraction

When light passes from one medium to another and undergoes a change in speed and direction, it is called refraction. The light does reflect but continues on its path with a change in its direction.

When light moves from a rarer to denser medium, i.e. moves from a light to heavy medium, the light tends to bend towards the normal while the opposite happens when light travels from a denser to rarer medium.

For the concept of refraction, one important thing to know is to learn about Snell's law.

This law tells us that,

$$\mu_1 \sin i = \mu_2 \sin r,$$

where μ_1 is the medium's refractive index from which light travels and μ_2 is the medium's refractive index to which light goes. The angle of incidence is i and the angle of refraction is r.

Generally, we deal with a situation where light goes from air to another medium and since the refractive index of air is 1, we get the equation,

$$\sin i = \mu \sin r$$

THEOREM

The lens formula is given by,

$$\frac{1}{v} - \frac{1}{u} = \frac{1}{f}.$$

Again it is important to use the sign conventions, or the answers are going to be wrong.

Most importantly, the sign conventions are going to remain the same as that of the mirrors, but here we have two focal lengths rather than one in the case of mirrors. So the focal length of concave is $-Ve$ by convention and that of convex is $+Ve$.

Example

If the image formed by a convex lens is $20\,\text{cm}$ from the lens and the radius of curvature is $30\,\text{cm}$ then find the distance of the object from the lens.

We have $R = 30\,\text{cm} \Rightarrow f = 15$. Using the lens formula, we have $\frac{1}{+20} - \frac{1}{u} = \frac{1}{15}$. Remember not to put the negative sign of u before solving it:

$$\frac{1}{20} - \frac{1}{15} = \frac{1}{u}$$

$$\frac{3-4}{60} = \frac{1}{u}$$

$$\frac{-1}{60} = \frac{1}{u}$$

$$u = -60\text{ cm}.$$

Theorem,

The lens formula we used above was a result of using a biconvex or biconcave lens. These lenses have equal radii of curvature and equal foci. However, the same cannot be used for a lens having different radii of curvature. So a new method of finding the focus, u and v was developed, called the lens maker formula.

$$= \left(\frac{\mu_2}{\mu_1} - 1\right)\left(\frac{1}{R_1} - \frac{1}{R_2}\right).$$

Here F is the resultant focal length, μ_2 is the refractive index of the object in which light enters, and μ_1 is the refractive index from which light travels to reach the object. Normally μ_1 is ignored when light travels from air because it has a refractive index of 1, so μ_1 becomes 1. R_1 is the radius of curvature of the 1st lens, and R_2 is the radius of curvature of the 2nd lens.

Atmospheric Refraction

We know that the earth is surrounded by a layer of air called the atmosphere. The density of air is not the same everywhere. The refractive index of air depends on its density, i.e. the higher the density of air, the greater its refractive index. The changes in refractive index can give rise to many phenomena.

For example, a rising current of hot air makes the objects viewed through it flicker and sway. You might have seen that on a summer day; hot air seems to be rising above the surface of the roads.

Early Sunrise and Late Sunset

Consider an oblique ray from a heavenly body such as the sun or a star. While travelling through the atmosphere, it continuously moves into regions of higher refractive index. So, it continuously bends towards the normal, resulting in a path similar to the figure given below.

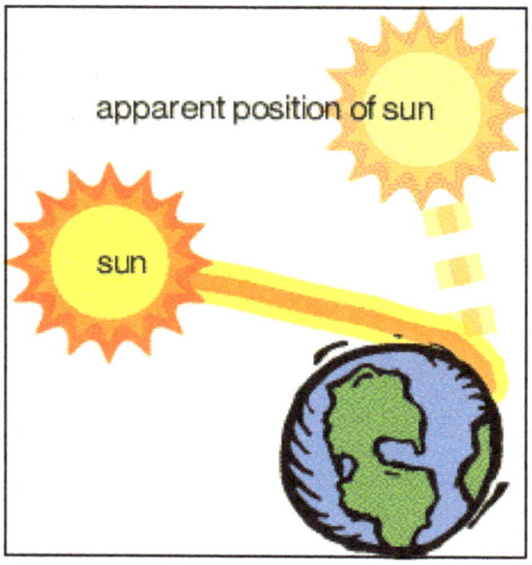

Since we see the object, that is the sun now, in the direction of the ray incident on the eye, the heavenly body appears higher than its actual position.

Now, consider a situation when the sun is just below the horizon. Rays of light coming from it get bent such that they seem to be coming from above the horizon (to the observer). Thus, when the sun's position is just below the horizon, the sun is visible to us. So, at sunrise, we see the sun before it actually comes to the horizon. At sunset, we see it even after it has dipped below the horizon. The fact is that it increases daylight by 4 minutes everyday (2 minutes at sunrise and 2 minutes at sunset).

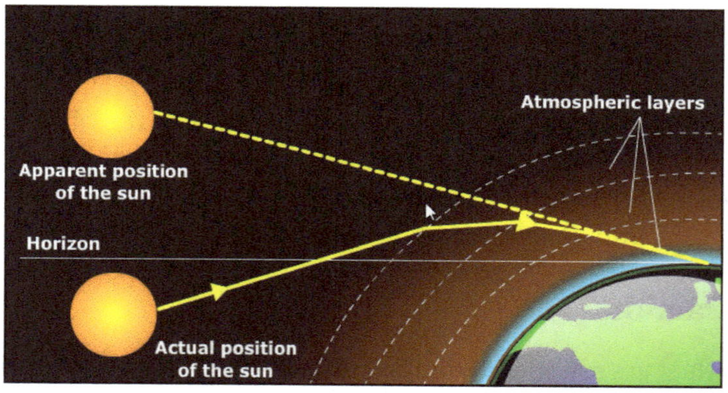

You might also have noticed that the sun looks oval/flattened at sunrise and sunset. The rays from the lower regions of the sun travel a bit greater distance than those rays which are from the middle and upper regions. So, they bend more. As a result, the image of the lower region gets shifted upwards more than that of the upper region. This makes the sun appear like a flattened circle or an oval.

Similarly, *twinkling of stars* is also a result of atmospheric refraction.

Now, we can define atmospheric refraction as follows:

Atmospheric refraction is a phenomenon, where light/any electromagnetic wave deviates from its rectilinear path because of passing through *atmosphere* where there are variations in the air density and refractive index.

Total Internal Reflection

When light travels from a denser to rarer medium with an angle greater than the critical angle, the ray of light does not deviate in its path or does not refract, but it undergoes a reflection known as total internal reflection.

The critical angle differs from medium to medium.

It is given by the formula,

$$\mu = \frac{1}{\sin \theta_c}.$$

This is very useful as it is used in fiber glasses where total internal reflection helps in fast movement of wavelengths.

Optical fibers are devices used for guiding light in many applications, most notably for fast communication. A fiber consists of a glass cylinder surrounded by a wall covered in a special coating.

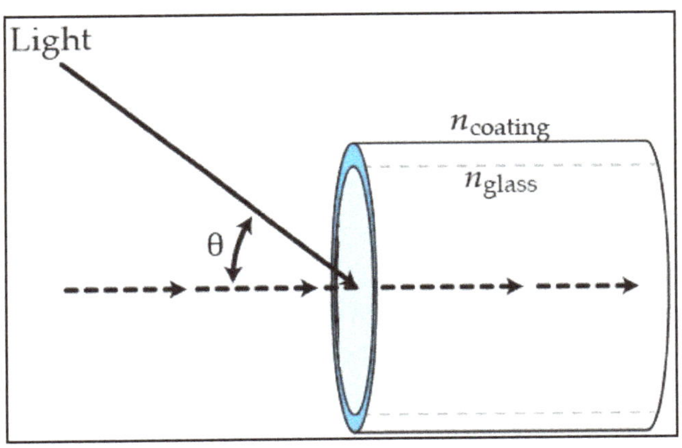

The fibers work on a principle called *total internal reflection*: light enters the fiber at an angle such that it does not get transmitted through the wall of the fiber when it hits the inside of the wall. Therefore, the refraction index of the glass part of the fiber has to be higher than that of its coating.

What is the maximum entering angle in degrees a light ray can pass from the air to the glass fiber for the total internal reflection to occur?

Details and Assumptions:

- Measure the entering angle from the axis of the fiber.
- Use the following refraction indexes: $n_{air} = 100$, $n_{glass} = 1.50$, and $n_{coating} = 1.46$.

Shifting of Image and Object

When viewing an object from different media the refraction of light often creates a shifting of the image. Because of this, for example, when viewing a pencil through a thick glass slab there is a shift of the image.

This shift of image is given by,

$$\mu = \frac{(\text{real height})}{(\text{apparent height})}$$

This is useful when viewing an object from water where the light bends only once.

However, when viewing an object from a glass slab, the light rays bend twice: once when going to the glass and again when coming out from the glass. So here we use

$$\Delta x = t\left(1 - \frac{1}{\mu}\right)$$

where Δx is the shift in the image and t is the thickness of the glass slab.

As a mathematical study, geometrical optics emerges as a short-wavelength limit for solutions to hyperbolic partial differential equations. In this short-wavelength limit, it is possible to approximate the solution locally by,

$$u(t,x) \approx a(t,x)e^{i(k \cdot x - \omega t)}$$

where k,ω satisfy a dispersion relation, and the amplitude $a(t,x)$ varies slowly. More precisely, the leading order solution takes the form,

$$a_0(t,x)e^{i\varphi(t,x)/\varepsilon}$$

The phase $\varphi(t,x)/\varepsilon$ can be linearized to recover large wavenumber $\varepsilon k := \nabla_x \varphi$, and frequency $\omega := -\partial_t \varphi$. The amplitude a_0 satisfies a transport equation. The small parameter ε enters the scene due to highly oscillatory initial conditions. Thus, when initial conditions oscillate much faster than the coefficients of the differential equation, solutions will be highly oscillatory, and transported along rays. Assuming coefficients in the differential equation are smooth, the rays will be too. In other words, refraction does not take place. The motivation for this technique comes from studying the typical scenario of light propagation where short wavelength light travels along rays that minimize (more or less) its travel time. Its full application requires tools from microlocal analysis.

A Simple Example

Starting with the wave equation for $(t,x) \in \mathbb{R} \times \mathbb{R}^n$

$$L(\partial_t, \nabla_x)u := \left(\frac{\partial^2}{\partial t^2} - c(x)^2 \Delta\right)u(t,x) = 0, \; u(0,x) = u_0(x), \; u_t(0,x) = 0$$

assume an asymptotic series solution of the form,

$$u(t,x) \sim a_\varepsilon(t,x)e^{i\varphi(t,x)/\varepsilon} = \sum_{j=0}^{\infty} i^j \varepsilon^j a_j(t,x)e^{i\varphi(t,x)/\varepsilon}.$$

Check that,

$$L(\partial_t,\nabla_x)(e^{i\varphi(t,x)/\varepsilon})a_\varepsilon(t,x) = e^{i\varphi(t,x)/\varepsilon}\left(\left(\frac{i}{\varepsilon}\right)^2 L(\varphi_t,\nabla_x\varphi)a_\varepsilon + \frac{2i}{\varepsilon}V(\partial_t,\nabla_x)a_\varepsilon + \frac{i}{\varepsilon}(a_\varepsilon L(\partial_t,\nabla_x)\varphi) + L(\partial_t,\nabla_x)a_\varepsilon\right)$$

with,

$$V(\partial_t,\nabla_x) := \frac{\partial\varphi}{\partial t}\frac{\partial}{\partial t} - c^2(x)\sum_j \frac{\partial\varphi}{\partial x_j}\frac{\partial}{\partial x_j}$$

Plugging the series into this equation, and equating powers of ε, the most singular term $O(\varepsilon^{-2})$ satisfies the eikonal equation (in this case called a dispersion relation),

$$0 = L(\varphi_t,\nabla_x\varphi) = (\varphi_t)^2 - c(x)^2(\nabla_x\varphi)^2.$$

To order ε^{-1}, the leading-order amplitude must satisfy a transport equation,

$$2Va_0 + (L\varphi)a_0 = 0$$

With the definition $k := \nabla_x\varphi$, $\omega := -\varphi_t$, the eikonal equation is precisely the dispersion relation that results by plugging the plane wave solution $e^{i(k\cdot x-\omega t)}$ into the wave equation. The value of this more complicated expansion is that plane waves cannot be solutions when the wave-speed c is non-constant. However, it can be shown that the amplitude a_0 and phase φ v are smooth, so that on a local scale there are plane waves.

To justify this technique, the remaining terms must be shown to be small in some sense. This can be done using energy estimates, and an assumption of rapidly oscillating initial conditions. It also must be shown that the series converges in some sense.

Reflection

"Reflection is a change in direction of a wave upon striking the interface between two materials."

When a ray of light strikes a plane mirror, the light ray reflects off the mirror. Reflection involves a change in direction of the light ray. The convention used to express the direction of a light ray is to indicate the angle which the light ray makes with a normal line drawn to the surface of the mirror. The angle of incidence is the angle between this normal line and the incident ray; the angle of reflection is the angle between this normal line and the reflected ray.

In optics, it relates to an image of an object.

Reflectional Symmetry

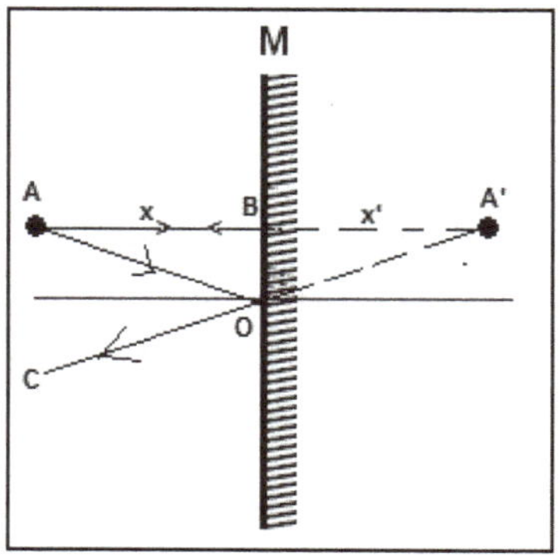

Look at the above diagram. An object A is placed in front of a mirror M. Consider rays AB and AO from A. As per law of reflection, the ray AB is reflected back as BA and ray AO reflected as OC.

Retracing these rays behind the mirror they appear to intersect at A'. This is where you find A', an image of A and it appears to be at the same distance behind as much as the object is before the mirror. That is the measures x and x' are same, though you cannot actually measure x'. The creation of image A' is actually called as the reflection of A.

Reflection of Light

Light rays are basically wave motions and they are capable of being reflected by an interface in between two mediums. Such interfaces are called Light Reflectors.

The most common Light reflector is the one we use many times a day, which is nothing but a plane mirror. But it is not only a plane mirror reflects light. Even a curved mirror does that though the nature of reflection of light by curved mirrors is different. Let us discuss this part separately at a later point.

Now for the present, let us concentrate on reflection of light by a plane mirror described by the following diagram.

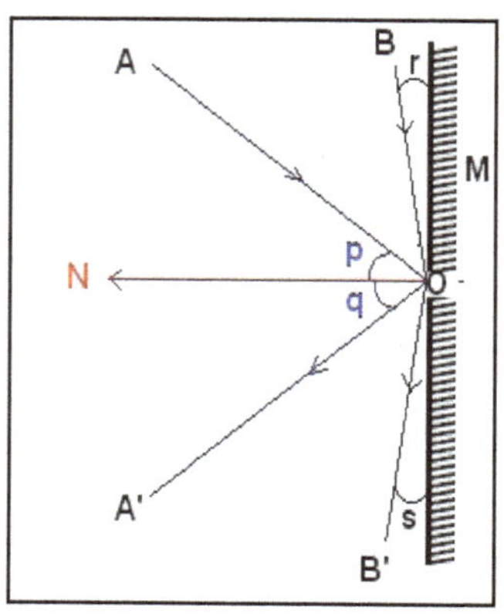

M is a plane mirror. It is a plane glass but on one side (the hatching show the surface) a suitable abrasive is coated to give the reflection effect to the glass, that is to make the glass to be a plane mirror. It also acts as an interface to separate the mediums on the left and right to that.

Let O be a any point on the plane mirror. The line NO is the perpendicular or better called as 'principal' to the plane mirror at O. Suppose a light ray AO hits the plane mirror at O, it is 'reflected to the same medium as ray OA'. This is called Reflection of light.

Types of Reflection

Diffuse Reflection

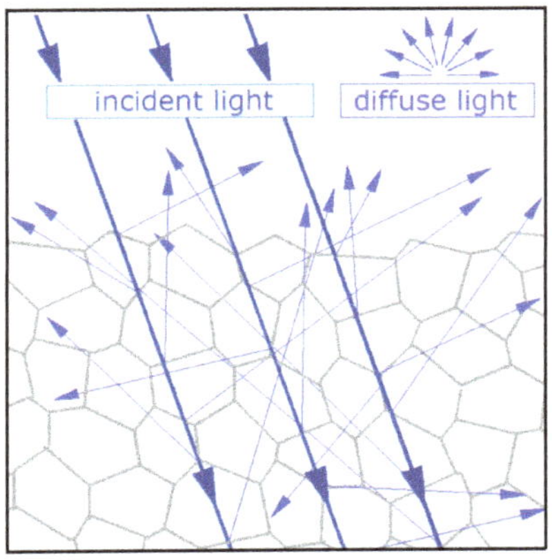

General scattering mechanism which gives diffuse reflection by a solid surface.

When light strikes the surface of a (non-metallic) material it bounces off in all directions due to multiple reflections by the microscopic irregularities *inside* the material (e.g. the grain boundaries of a polycrystalline material, or the cellor fiber boundaries of an organic material) and by its surface, if it is rough. Thus, an ‹image' is not formed. This is called *diffuse reflection*. The exact form of the reflection depends on the structure of the material. One common model for diffuse reflection is Lambertian reflectance, in which the light is reflected with equal luminance (in photometry) or radiance (in radiometry) in all directions, as defined by Lambert's cosine law.

The light sent to our eyes by most of the objects we see is due to diffuse reflection from their surface, so that this is our primary mechanism of physical observation.

Retroreflection

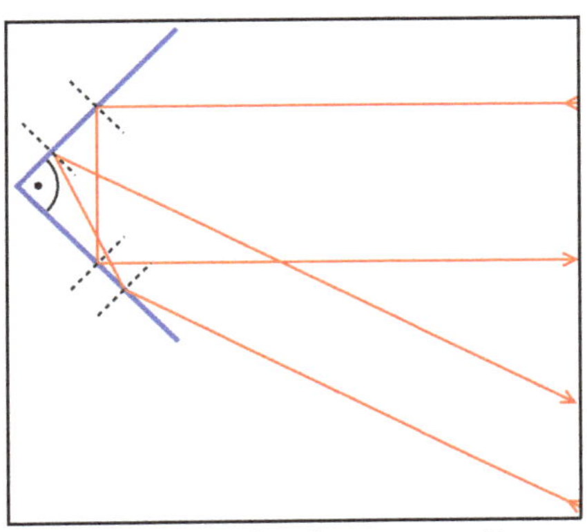

Working principle of a corner reflector.

Some surfaces exhibit *retroreflection*. The structure of these surfaces is such that light is returned in the direction from which it came.

When flying over clouds illuminated by sunlight the region seen around the aircraft's shadow will appear brighter, and a similar effect may be seen from dew on grass. This partial retro-reflection is created by the refractive properties of the curved droplet's surface and reflective properties at the backside of the droplet.

Some animals' retinas act as retroreflectors as this effectively improves the animals' night vision. Since the lenses of their eyes modify reciprocally the paths of the incoming and outgoing light the effect is that the eyes act as a strong retroreflector, sometimes seen at night when walking in wildlands with a flashlight.

A simple retroreflector can be made by placing three ordinary mirrors mutually perpendicular to one another (a corner reflector). The image produced is the inverse of one produced by a single mirror. A surface can be made partially retroreflective by depositing a layer of tiny refractive spheres on it or by creating small pyramid like structures. In both cases internal reflection causes the light to be reflected back to where it originated. This is used to make traffic signs and automobile license plates reflect light mostly back in the direction from which it came. In this application perfect retroreflection is not desired, since the light would then be directed back into the headlights of an oncoming car rather than to the driver's eyes.

Multiple Reflections

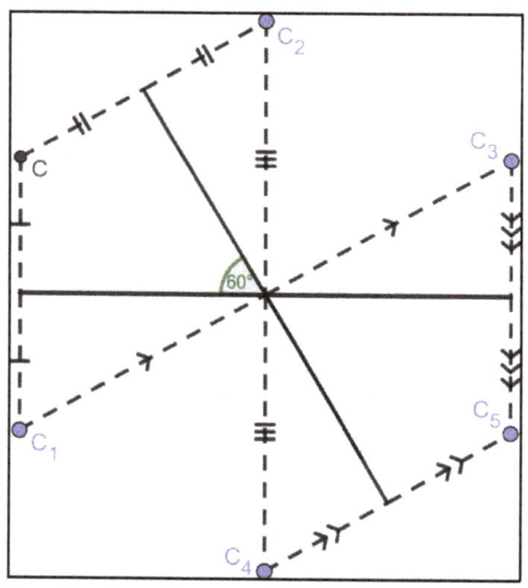

Multiple reflections in two plane mirrors at a 60° angle.

When light reflects off a mirror, one image appears. Two mirrors placed exactly face to face give the appearance of an infinite number of images along a straight line. The multiple images seen between two mirrors that sit at an angle to each other lie over a circle. The center of that circle is located at the imaginary intersection of the mirrors. A square of four mirrors placed face to face give the appearance of an infinite number of images arranged in a plane. The multiple images seen between four mirrors assembling a pyramid, in which each pair of mirrors sits an angle to each other, lie over a sphere. If the base of the pyramid is rectangle shaped, the images spread over a section of a torus.

Note that these are theoretical ideals, requiring perfect alignment of perfectly smooth, perfectly flat perfect reflectors that absorb none of the light. In practice, these situations can only be approached but not achieved because the effects of any surface imperfections in the reflectors propagate and magnify, absorption gradually extinguishes the image, and any observing equipment (biological or technological) will interfere.

Complex Conjugate Reflection

In this process (which is also known as phase conjugation), light bounces exactly back in the direction from which it came due to a nonlinear optical process. Not only the direction of the light is reversed, but the actual wavefronts are reversed as well. A conjugate reflector can be used to remove aberrations from a beam by reflecting it and then passing the reflection through the aberrating optics a second time.

Other Types of Reflection

Neutron Reflection

Materials that reflect neutrons, for example beryllium, are used in nuclear reactors and nuclear weapons. In the physical and biological sciences, the reflection of neutrons off of atoms within a material is commonly used to determine the material's internal structure.

Sound Reflection

Sound diffusion panel for high frequencies.

When a longitudinal sound wave strikes a flat surface, sound is reflected in a coherent manner provided that the dimension of the reflective surface is large compared to the wavelength of the sound. Note that audible sound has a very wide frequency range (from 20 to about 17000 Hz), and thus a very wide range of wavelengths (from about 20 mm to 17 m). As a result, the overall nature of the reflection varies according to the texture and structure of the surface. For example, porous materials will absorb some energy, and rough materials (where rough is relative to the wavelength) tend to reflect in many directions—to scatter the energy, rather than to reflect it coherently. This leads into the field of architectural acoustics, because the nature of these reflections is critical to the auditory feel of a space. In the theory of exterior noise mitigation, reflective surface size mildly detracts from the concept of a noise barrier by reflecting some of the sound into the opposite direction. Sound reflection can affect the acoustic space.

Seismic Reflection

Seismic waves produced by earthquakes or other sources (such as explosions) may be reflected by layers within the Earth. Study of the deep reflections of waves generated by earthquakes has allowed seismologists to determine the layered structure of the Earth. Shallower reflections are used in reflection seismology to study the Earth's crust generally, and in particular to prospect for petroleum and natural gas deposits.

Law of Reflection

Let us first study what is law of reflection with respect to a single light ray. Look at the following diagram.

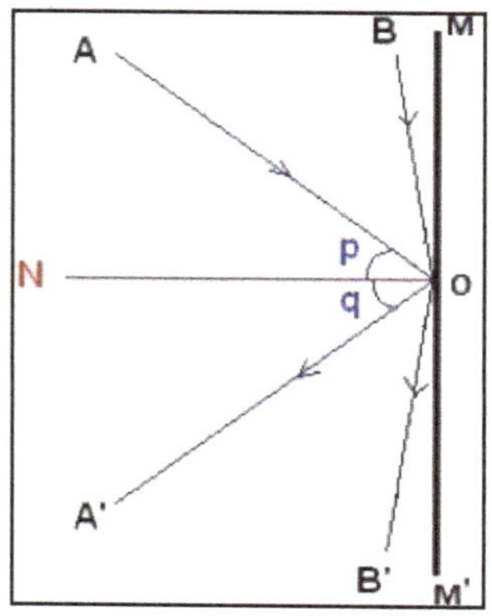

Let MM' be a plane mirror which is absolutely smooth. A light ray AO strikes the mirror at O (the incident ray), and it is reflected as ray OA', called as reflected ray. ON is the perpendicular to MM' at O. It is also called as the normal to the surface at the point of reflection.

The angle made by the incident ray with the normal ON is p and q be the angle made by the reflected ray with the same normal. These are called as 'Angle of Incidence' and the 'Angle of Reflection' respectively.

According to the Law of Reflection, Angle of incidence is always equal to angle of reflection.

Law of Reflection Definition

Now, let us consider a set of rays assuming a plane mirror with a smooth and perfect surface. The situation is explained in the following diagram:

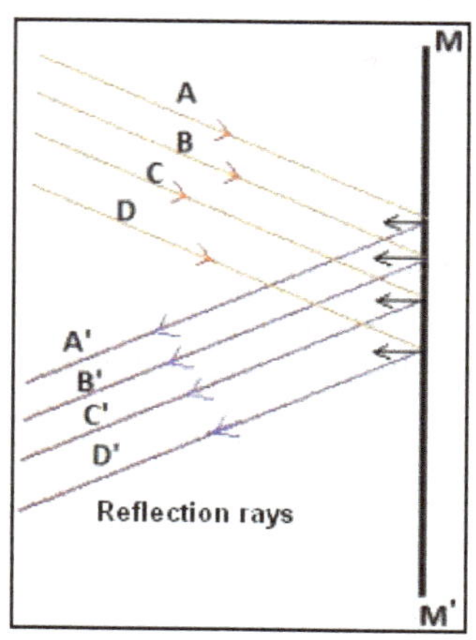

Let ABCD is a set of rays striking a mirror MM'. Obviously, the rays of parallel to each other. When each individual ray strikes the mirror it is reflected as per law of reflection, that is the angle of incidence and the angle of reflection are same for each ray. Thus the set of incident rays are reflected as a set of rays. In other words, a beam of light is reflected as a beam of light if the mirror is plane and its reflecting surface is smooth. This type of reflection of a set of rays is called Specular Reflection.

Thus, observing at this diagram we can Define Law of Reflection as:

When light falls on a surface which is plane, the angle with which the light falls on the surface is equal to the angle with which the light reflects back and the incident ray, reflected ray and the normal ray lies in the same plane.

Does the Law of Reflection Hold for Diffuse Reflection?

What happens if the reflecting surface is rough or curved? Does the law of reflection hold for diffuse reflection? Does the law of reflection hold for curved mirrors? Diffused Reflection takes place when the incident light falls on the rough surface.

Let us study the situation again considering each ray in a set of rays through the following diagram.

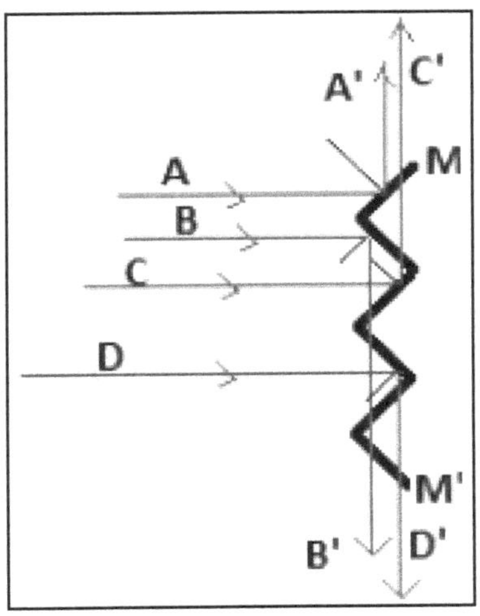

Consider a reflecting surface MM' which is highly exaggerated rough for the purpose of study. A set of rays ABCD strike the reflecting surface. But when you consider the rays individually, the infinitesimally small reflecting surface on which each ray falls is plane. Hence as far as that ray is concerned the law of reflection holds good and the reflection obeys the law that angle of incidence is equal to the angle of reflection. For example, the ray A is reflected as A'. Overall, the set of rays ABCD is reflected as A'B'C'D' as shown in the diagram.

It may be noticed that while the incident rays are parallel, the reflected rays emerge in different directions. This is called scatter reflection or diffuse reflection. Thus the law of reflection still holds in case of diffuse reflection but the net is result different from specular reflection.

Does the Law of Reflection Hold for Curved Mirrors?

Now let us consider the case of a curved mirror. As an example a convex mirror is shown in the following diagram.

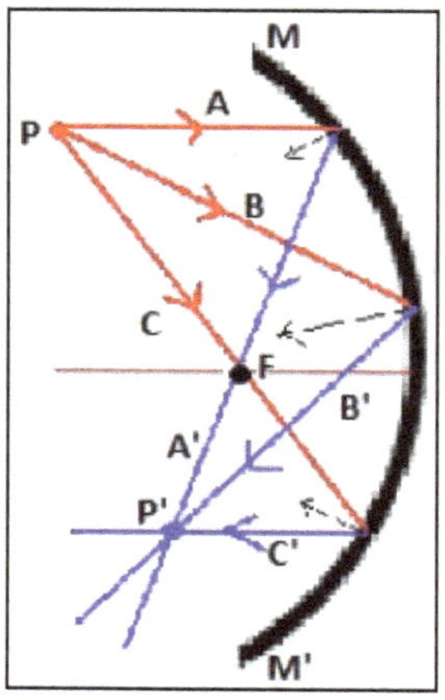

MM' is a concave mirror and rays A, B and C from an object P emerging out in different directions. Each of these is reflected following the law of reflection at the respective striking points. Thus the law of reflection holds good for curved mirrors also.

The corresponding reflected rays are A', B' and C' respectively and we find these rays meet at a point P' to create an image of the object P.

An incident ray, that is parallel to the principal axis of the mirror is reflected through a point F on the principal axis and an incident ray through F is reflected parallel to the principal axis. This is called two rules of a curved mirror and simplifies the job of drawing a ray diagram. The point F is called the Focal point of the mirror.

Hence, we can Conclude that Law of reflection does not hold for Curved mirrors as the angle of incidence is different from the angle of reflection.

Angle of Incidence

The Angle of incidence can be defined as the angle with which a light ray strikes a reflecting surface. But the angle of strike is measured as the angle between the ray and the normal to the surface at the point of incidence of the ray. It has a relation with the angle of reflection and also with the type of reflecting surface.

Angle of Incidence Definition

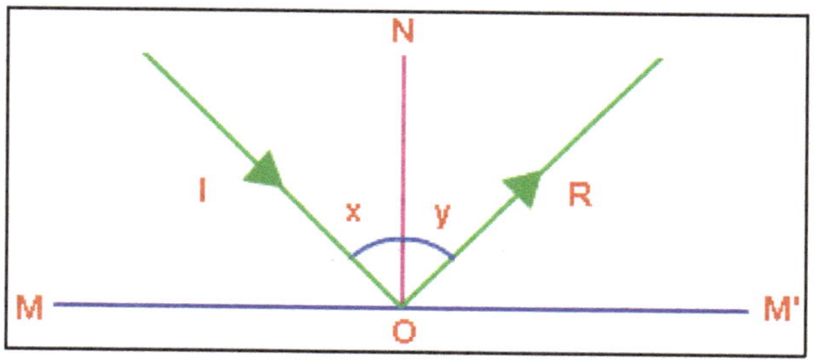

The ray I from a light source strikes the mirror MM' at point O This ray is called as incident ray. ON is the perpendicular to the mirror MM' at O which is also called the normal to the mirror at O. The angle 'x' made by the ray I, with the normal ON is called the 'angle of incidence'.

The incident ray, is reflected by the mirror when it strikes the mirror at O and the ray after reflection 'R' is called as 'reflected ray'. The angle 'y' made by the reflected ray with the same ray is called 'angle of reflection'.

Angle of Incidence and Reflection

As per law of reflection the angle of incidence is always equal to angle of reflection. This can be proved mathematically and also by a simple experiment. There are many mathematical proofs but the simplest proof is based on Fermat's Principle.

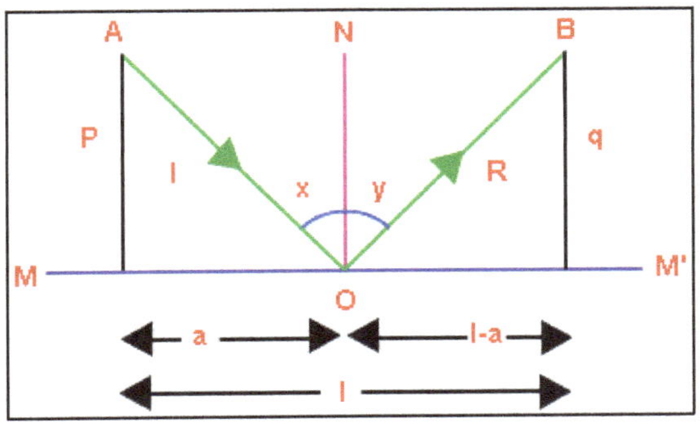

Look at the above diagram. Let a right ray AO from a source A strike a plane mirror at O and reflected as ray OB. Let B be a point on ray OB and let ON be the normal to the mirror at O. The angle of incidence is 'x' and angle of reflection is 'y', as shown. Let 't' be the time taken for the ray to travel the distance AO + OB and 'c' be the velocity of light, which is a constant.

The time 't' can be worked out as,

$$t = \frac{AO}{c} + \frac{OB}{c} \text{ or,}$$

$$t = \frac{\sqrt{a^2 + p^2}}{c} + \frac{\sqrt{(l-a)^2 + q^2}}{c}$$

As per Fermat's principle light takes the path of least time. Therefore, the derivative $\frac{dt}{da}$ must be 0.

$$\text{Thus is } 0 = \frac{1}{c}\left[\frac{a}{\sqrt{a^2 + p^2}} + \frac{-(l-a)}{\sqrt{(l-a)^2 + q^2}}\right]$$

$$\text{or, } \frac{a}{\sqrt{a^2 + p^2}} = \frac{-(l-a)}{\sqrt{(l-a)^2 + q^2}}$$

But as per trigonometry, this is same as $sin(x) = sin(y)$ or $x = y$.

Hence,

The Angle of Incidence = Angle of Reflection

The law of reflection can also be proved experimentally. The arrangement for the experiment is described below.

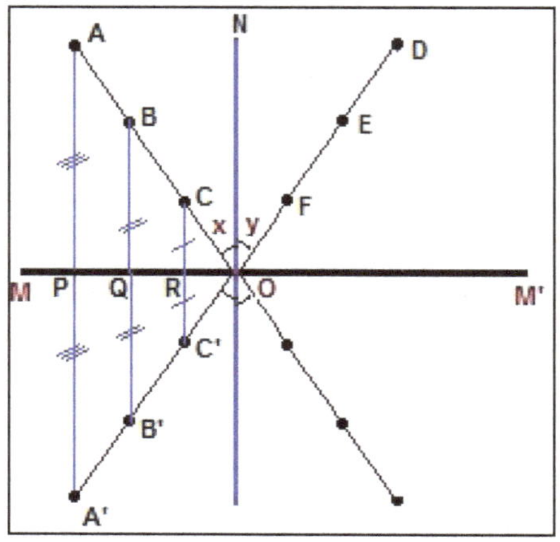

Take a white card board sheet and fix the same firmly on a board. Draw a thick line MM' and mark the midpoint O. Draw a perpendicular ON at O. Place a plain mirror on MM' with its center at O. Take three pins and fix them vertically on the sheet at positions C, B and A on one side of ON such that the pins lie on a straight line when viewed from A. Now move to the other side of the perpendicular ON and fix three more pins at F, E and D in such way that the images A' B' C' of the pins at A, B, C and the pins at D, E, F are all in one straight line. This is the most important point of the entire experiment.

Now remove the mirror and the pins carefully marking the points where they have been fixed. At this stage we will not be able to locate the positions of A', B' and C'. Draw a perpendicular AP from A and extend that line. Join DO and extend that too. Let A' be the point of intersection of these extended line. A' is nothing but the position of the image that appeared when the mirror was there. Same way locate the positions of B' and C'.

If you join AO and measure the angles AON and DON, you will find them to be equal. Thus, it proves the angle of incidence equals angle of reflection. Also you will notice that the following measures to be equal. AP = PA', BQ = QB' and CR = RC', which proves the image distance equals the object distance.

Angle of Reflection

Referring to the same diagram, the angle A'ON between the reflected ray OA' and the principal axis NO is called as 'Angle of Reflection'.

The Angle of reflection is the angular reflection formed by a reflected ray and a perpendicular to the surface at the point of reflection. The Angle of Reflection depends on the Angle of Incidence.

Angle of Reflection Formula

The Angle of incidence always equals the angle of reflection and the distance of image always equals the distance of object are called Angle of reflection formula.

if θ_i and θ_r are the angle of incidence and angle of reflection respectively, then

$$\sin\theta_i = \sin\theta_r$$

or

$$\theta_i = \theta_r$$

This is the Angle of Reflection Formula.

Grazing Angle

Now, in the same diagram, we see another ray BO which falls on the mirror and reflected as as a ray OB'. In this case, it is easier to perceive the angles at which the incident ray and the reflected ray with the surface of the mirror rather than to its normal. They are angles 'r' and 's' and these are called 'grazing angles'. The angles 'r' and's' are congruent because of the fact they are the difference of 90° – the angle of incidence or reflection as the case may be.

Specular Reflection

Specular reflection is a type of surface reflectance often described as a mirror-like reflection of light from the surface. In specular reflection, the incident light is reflected into a single outgoing direction. The name specular is derived from the Latin word speculum, meaning mirror.

Specular reflection obeys the Laws of Reflection and the light is reflected from a smooth surface at a definite angle. Specular reflection occurs for all polished and smooth surfaces, like mirrors. The angle of reflection in specular reflection is equal to the angle of incidence. Rough or unpolished surfaces are not good specular reflectors as they do not reflect light at the same angle as the angle of incidence.

Mirrors that are used for specular reflection are usually made of metals such as aluminum or silver. Specular reflection is measured using incidence reflectometers by scanning a variable wavelength source.

If a specular reflection light ray goes straight into a camera, it is as if the camera is seeing the light source directly, even though it has bounced off of an object. The camera is seeing the light of the light source, not the color of the object. If you have a white light source, the specular reflection will be white. If you have a red light source, the specular reflection will be red. Therefore, to model specular reflection you need to specify the color of the light source in your lighting model.

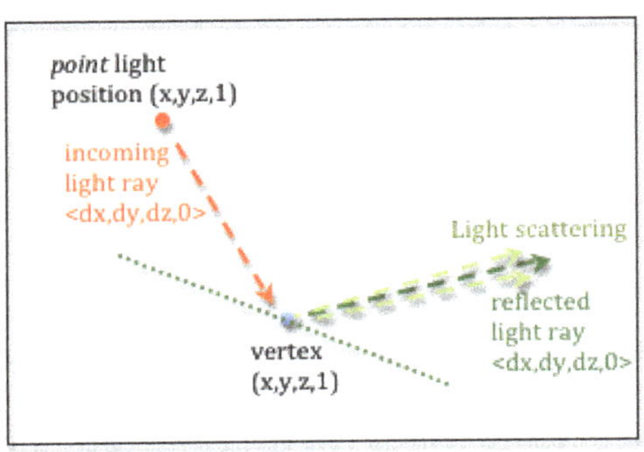

Specular light scattering.

We assume there is a small amount of light scattering around the reflection vector, but not much. The scattering is centered around the reflection vector as shown in the diagram to the left.

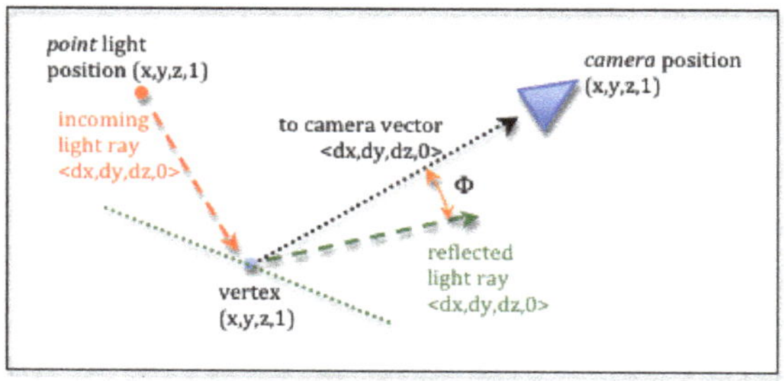

Specular light intensity.

If we calculate a vector from a fragment to the camera, and then calculate the angle between it and the reflection vector, we can use the angle to estimate the amount of specular reflection the camera receives. If the angle is zero the camera is receiving the entire reflection light ray. If the angle is very small, then some percentage of the reflection ray is going into the camera. If the angle is large, then no light from the reflection is entering the camera. If the angle is between small and large, we need a way to calculate some percentage of reflected light that is entering the camera. We used a cosine function to calculate diffuse light percentages, but the cosine function is too broad for the a focused specular reflection. However, if we raise the cosine function to a power, the curve collapses around the Y axis. Experiment with various exponents in the plot of the $\cos(\text{angle})^{\text{exp}}$ function below. By using various exponents for this equation, we can simulate various amounts of light scattering around the specular reflection vector. If the exponent is large, such as 100, then the $\cos(\text{angle})^{\text{exp}}$ shrinks around the Y axis and only very small angles will return a significant percentage value. If the exponent is small, such as 1.0, a broad amount of light around the reflection ray will be simulated.

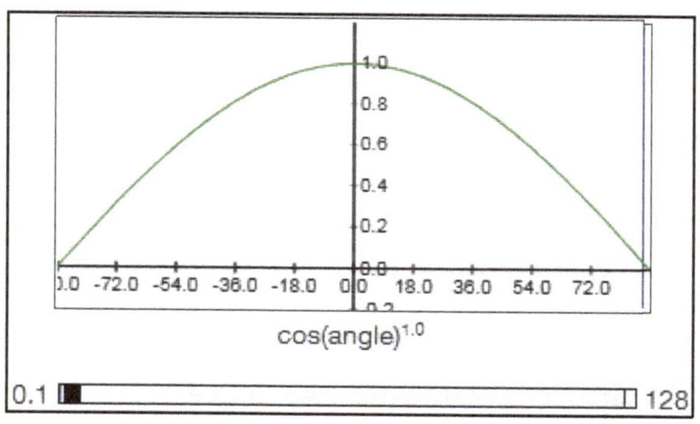

The Math for Specular Reflection

To calculate specular reflection, we need two vectors:

- A vector from the fragment location to the camera.
- A reflected light vector.

The angle between these two vectors determines the amount of specular reflection.

To calculate a vector from the fragment position to the camera, subtract the head of the vector (the camera) from the tail (the fragment location). As we previously discussed, we are performing these calculations in "camera space" and the camera is located at the global origin, (0,0,0). Therefore, a vector to the camera is:

$$to\_camera[0] = 0 - fragment\_position[0]$$
$$to\_camera[1] = 0 - fragment\_position[1]$$
$$to\_camera[2] = 0 - fragment\_position[2]$$

We can calculate this vector by simply multiplying the fragment location by -1.

Calculating the Reflection Vector

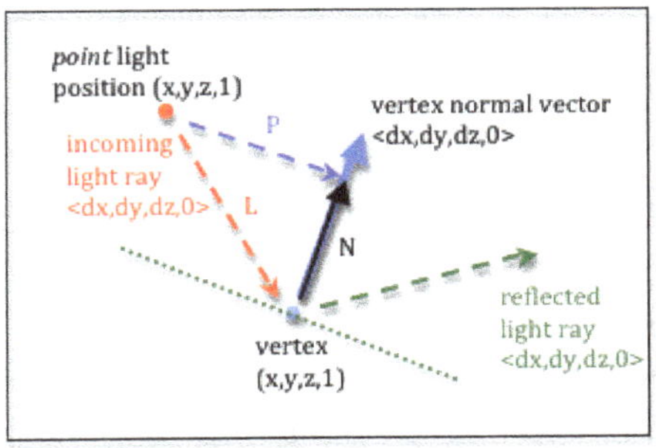

Calculating the reflection vector

To calculate the reflection vector, we need to set up several intermediate vectors. Please study the diagram to the right before proceeding with this discussion. The position of the light source can be projected onto the vertex's normal vector. This is the "p" vector in the diagram. It can be shown that a vector in the direction of the normal vector that has a length to the projected point can be calculated by taking the dot product of -L and a normalized vertex normal vector. The dot product calculates the angle between the two vectors, but only when the vectors have unit length. Notice that the normal vector is normalized, but the -L vector is not. This provides the correct length for the N vector to the projected point. To summarize, to calculate N we:

- Normalize the vertex's normal vector to make it one unit in length. Let's call it n.
- Calculate the dot product of -L and n. Let's call this value *s*.
- Scale n by *s* to create the vector N, which is in the direction of the vertex's normal vector and has a length to the projected point of the light source.
- From simple vector addition, L + N === P.

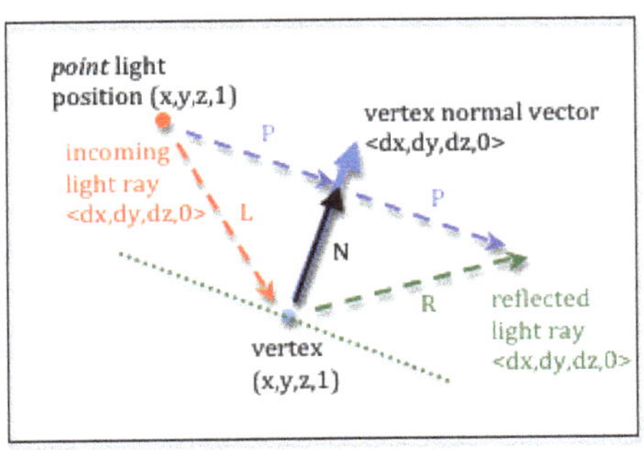

Calculating the reflection vector.

Now we know the vector P. Observe that from simple vector addition, the reflection vector R is equal to N + P. We can substitute the calculations for N and P into this equation like this:

$$R = N + PR$$
$$= n * dot\_product(n,-L) + (L + N)$$
$$R = n * dot\_product(n,-L) + (L + n * dot\_product(n,-L))$$
$$R = 2 * n * dot\_product(n,-L) + L$$

If you flip the direction of L and follow the same logic, you get:

$$R = 2 * n * dot\_product(n,L) - L \; // \; \textit{When L goes from the vertex to the light sourc}$$

The Percentage of a Reflection Ray a Camera Receives

We have a vector from the vertex to the camera and we have a reflection vector. The angle between these two vectors indicates how much of the reflection ray goes into the camera. The cosine of this angle, which is a percentage between 0.0 and 1.0 is raised to some power to simulate a focused bean of light around the reflection ray. The exponent is typically called the “shininess” exponent. The larger the exponent, the more shiny an object appears. Shiny objects have small, focused specular reflection. Dull objects have large, spread out specular reflection.

Specular Reflection in Shader Programs

Please study the following *shader programs*. Then compare the programs to the comments below.

Vertex Shader

```
// Vertex Shader
precision mediump int;
precision mediump float;

// Scene transformations
uniform mat4 u_PVM_transform; // Projection, view, model transform
uniform mat4 u_VM_transform;  // View, model transform

// Light model
uniform vec3 u_Light_position;
uniform vec3 u_Light_color;
uniform float u_Shininess;
```

```
// Original model data
attribute vec3 a_Vertex;
attribute vec3 a_Color;
attribute vec3 a_Vertex_normal;

// Data (to be interpolated) that is passed on to the fragment shader
varying vec3 v_Vertex;
varying vec4 v_Color;
varying vec3 v_Normal;

void main() {

  // Perform the model and view transformations on the vertex and pass
this
  // location to the fragment shader.
  v_Vertex = vec3( u_VM_transform * vec4(a_Vertex, 1.0) );

  // Perform the model and view transformations on the vertex's normal
vector
  // and pass this normal vector to the fragment shader.
  v_Normal = vec3( u_VM_transform * vec4(a_Vertex_normal, 0.0) );

  // Pass the vertex's color to the fragment shader.
  v_Color = vec4(a_Color, 1.0);

  // Transform the location of the vertex for the rest of the graphics
pipeline
  gl_Position = u_PVM_transform * vec4(a_Vertex, 1.0);
}
```

- This is the same vertex shader program used for diffuse lighting in the previous lesson, with the addition of two new uniform variables: u_Light_color and u_Shininess. These values enhance our lighting model.

- All calculations in the *fragment shader* will be done in "camera space," so the vertex data is transformed by the *model* and *camera* transformations, but not the *projection* transformation.

Fragment Shader

The fragment shader calculates a reflection vector and then determines if any reflected light should be used to color the pixel.

```
// Fragment shader program
precision mediump int;
precision mediump float;

// Light model
uniform vec3 u_Light_position;
uniform vec3 u_Light_color;
uniform float u_Shininess;

// Data coming from the vertex shader
varying vec3 v_Vertex;
varying vec4 v_Color;
varying vec3 v_Normal;

void main() {
  vec3 to_light;
  vec3 vertex_normal;
  vec3 reflection;
  vec3 to_camera;
  float cos_angle;
  vec3 specular_color;
  vec3 object_color;
  vec3 color;

  // Calculate a vector from the fragment location to the light source
  to_light = u_Light_position - v_Vertex;
```

```
  to_light = normalize( to_light );

  // The vertex's normal vector is being interpolated across the primi-
tive
  // which can make it un-normalized. So normalize the vertex's normal
vector.
  vertex_normal = normalize( v_Normal );

  // Calculate the reflection vector
  reflection = 2.0 * dot(vertex_normal,to_light) * vertex_normal - to_
light;

  // Calculate a vector from the fragment location to the camera.
  // The camera is at the origin, so negating the vertex location gives
the vector
  to_camera = -1.0 * v_Vertex;

  // Calculate the cosine of the angle between the reflection vector
  // and the vector going to the camera.
  reflection = normalize( reflection );
  to_camera = normalize( to_camera );
  cos_angle = dot(reflection, to_camera);
  cos_angle = clamp(cos_angle, 0.0, 1.0);
  cos_angle = pow(cos_angle, u_Shininess);

  // If this fragment gets a specular reflection, use the light's color,
  // otherwise use the objects's color
  specular_color = u_Light_color * cos_angle;
  object_color = vec3(v_Color) * (1.0 - cos_angle);
  color = specular_color + object_color;

  gl_FragColor = vec4(color, v_Color.a);
}
```

- The angle between a vector to the camera and a reflection vector is calculated. The calculations are exactly as described in the math section above.
- It is important to note that the cosine of the angle is clamped to a value between 0.0 and 1.0 before is it raised to the shininess exponent. The angle can be negative if the reflection vector and the vector to the camera have an angle greater than 90 degrees. Raising a number to a power can make a value positive. For example, $(-2)^2$ is +4. (If the cosine of the angle is negative, it will be clamped to zero, which means 0% of reflected light gets to the camera, which is the physical result we want.)
- If the value of the cos(angle) is one, we want the color of the light source for this fragment. If the value of the cos(angle) is zero, we want the color of the object. If the cos(angle) is some percentage between zero and one, we take this percentage of the light's color, and the left-over percentage, (1.0 - cos_angle), of the color of the object. "Color is additive" so we can simply add the two color values together.

Law of Reflection

The law of reflection describes the angle of reflected light: the angle of incident light is the same as the angle of the reflected light.

The law of reflection arises from diffraction of a plane wave with small wavelength on a flat boundary: when the boundary size is much larger than the wavelength, then electrons of the boundary are seen oscillating exactly in phase only from one direction – the specular direction. If a mirror becomes very small compared to the wavelength, the law of reflection no longer holds, and the behavior of light is more complicated.

Vector Formulation

The law of reflection can also be equivalently expressed using linear algebra. The direction of a reflected ray is determined by the vector of incidence and the surface normal vector. Given an incident direction $\hat{\mathbf{d}}_\mathrm{i}$ from the surface to the light source and the surface normal direction $\hat{\mathbf{d}}_\mathrm{n}$ the specularly reflected direction $\hat{\mathbf{d}}_\mathrm{s}$ (all unit vectors) is:

$$\hat{\mathbf{d}}_\mathrm{s} = 2\left(\hat{\mathbf{d}}_\mathrm{n} \cdot \hat{\mathbf{d}}_\mathrm{i}\right)\hat{\mathbf{d}}_\mathrm{n} - \hat{\mathbf{d}}_\mathrm{i},$$

where $\hat{\mathbf{d}}_\mathrm{n} \cdot \hat{\mathbf{d}}_\mathrm{i}$ is a scalar obtained with the dot product. Different authors may define the incident and reflection directions with different signs. Assuming these Euclidean vectors are represented in column form, the equation can be equivalently expressed as a matrix-vector multiplication:

$$\hat{\mathbf{d}}_\mathrm{s} = \mathbf{R}\,\hat{\mathbf{d}}_\mathrm{i},$$

where $\mathbf{R}$ is the so-called Householder transformation matrix, defined as:

$$\mathbf{R} = \mathbf{I} - 2\hat{\mathbf{d}}_\mathrm{n}\hat{\mathbf{d}}_\mathrm{n}^\mathrm{T};$$

T denotes transposition and $\mathbf{I}$ is the identity matrix.

Reflectivity

Reflectivity is the ratio of the power of the reflected wave to that of the incident wave. It is a function of the wavelength of radiation, and is related to the refractive index of the material as expressed by Fresnel's equations. In regions of the electromagnetic spectrum in which absorption by the material is significant, it is related to the electronic absorption spectrum through the imaginary component of the complex refractive index. The electronic absorption spectrum of an opaque material, which is difficult or impossible to measure directly, may therefore be indirectly determined from the reflection spectrum by a Kramers-Kronig transform. The polarization of the reflected light depends on the symmetry of the arrangement of the incident probing light with respect to the absorbing transitions dipole moments in the material.

Measurement of specular reflection is performed with normal or varying incidence reflection spectrophotometers (*reflectometer*) using a scanning variable-wavelength light source. Lower quality measurements using a glossmeter quantify the glossy appearance of a surface in gloss units.

Consequences

Internal Reflection

When light is propagating in a material and strikes an interface with a material of lower index of refraction, some of the light is reflected. If the angle of incidence is greater than the critical angle, total internal reflection occurs: all of the light is reflected. The critical angle can be shown to be given by,

$$\theta_{crit} = \arcsin\left(\frac{n_2}{n_1}\right).$$

Polarization

When light strikes an interface between two materials, the reflected light is generally partially polarized. However, if the light strikes the interface at Brewster's angle, the reflected light is *completely* linearly polarized parallel to the interface. Brewster's angle is given by,

$$\theta_B = \arctan\left(\frac{n_2}{n_1}\right).$$

Reflected Images

The image in a flat mirror has these features:

- It is the same distance behind the mirror as the object is in front.
- It is the same size as the object.
- It is the right way up (erect).
- It is reversed.
- It is *virtual*, meaning that the image appears to be behind the mirror, and cannot be projected onto a screen.

The reversal of images by a plane mirror is perceived differently depending on the circumstances. In many cases, the image in a mirror appears to be reversed from left to right. If a flat mirror is mounted on the ceiling it can appear to reverse *up* and *down* if a person stands under it and looks up at it. Similarly a car turning *left* will still appear to be turning *left* in the rear view mirror for the driver of a car in front of it. The reversal of directions, or lack thereof, depends on how the directions are defined. More specifically a mirror changes the handedness of the coordinate system, one axis of the coordinate system appears to be reversed, and the chirality of the image may change. For example, the image of a right shoe will look like a left shoe.

Examples

Esplanade of the Trocadero in Paris after rain. The layer of water exhibits specular reflection, reflecting an image of the Eiffel Tower and other objects.

A classic example of specular reflection is a mirror, which is specifically designed for specular reflection.

In addition to visible light, specular reflection can be observed in the ionospheric reflection of radiowaves and the reflection of radio- or microwave radar signals by flying objects. The measurement technique of x-ray reflectivityexploits specular reflectivity to study thin films and interfaces with sub-nanometer resolution, using either modern laboratory sources or synchrotron x-rays.

Non-electromagnetic waves can also exhibit specular reflection, as in acoustic mirrors which reflect sound, and atomic mirrors, which reflect neutral atoms. For the efficient reflection of atoms from a solid-state mirror, very cold atoms and/or grazing incidence are used in order to provide significant quantum reflection; ridged mirrors are used to enhance the specular reflection of atoms. Neutron reflectometry utilizes specular reflection to study material surfaces and thin film interfaces in an analogous fashion to x-ray reflectivity.

Diffuse Reflection

Diffuse reflection is the property of light to scatter when reflecting off a surface. Light travels in a straight line, and when it strikes an object, its expected primary angle of reflection is called a specular reflection. In addition, the light also reflects in all possible angles and directions. The degree or magnitude of this diffused reflection is dependent upon the characteristics of the reflecting substance and surface.

Some surfaces, such as a highly polished metal mirror, reflect light at nearly 100% specular efficiency. Other surfaces, such as those of some crystals or liquids, may most or all of it to pass through the surface and medium. Most objects of the world, including the above examples, reflect light both specularly and diffusely to varying degrees. Diffuse reflection is the primary property of light that allows human eyes to see an object.

The main determinant of reflection is the surface's absorption of light. Polished surfaces such as white marble stone, or the randomly irregular fibers of paper, reflect diffuse light with nearly equal efficiency. Black objects tend to absorb more light. Some naturally occurring as well as human made things emit light and overwhelm any diffuse reflection from their surface to distinguish them as objects.

Diffuse Reflection of Ambient Light

If we assume a uniform intensity of ambient light, Ia, then the intensity of the diffuse reflection at any point on the surface: I = ka * Ia, where ka = the ambient diffuse coefficient of reflection with: 0.0 <= ka <= 1.0. For colored surfaces there are three ka values, one for red, green, and blue. We are approximating the ambient diffuse part of the reflection intensity function R(j,q,l) by the constant ka. For a dull, non-reflecting surface ka is close to 0.0 and for a highly reflective surface, such as a mirror, ka is close to 1.0.

Diffuse Reflection of Point Light Sources

Consider a point source and assume the source is far from the surface so that incident light rays are parallel. Note: if the light source is near the surface then the light rays are not parallel as below:

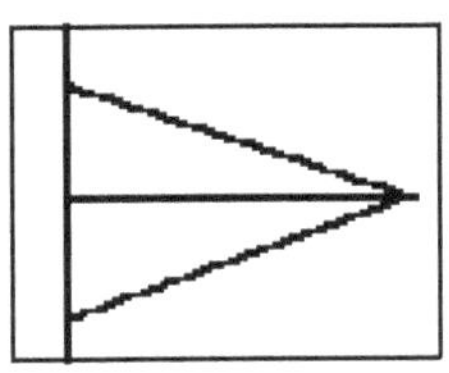

For a point light far from the surface, we use Lambert's cosine law which is as follows: the intensity of reflected light is f(angle of illumination). This can be qualitatively seen if we look at the two cases below:

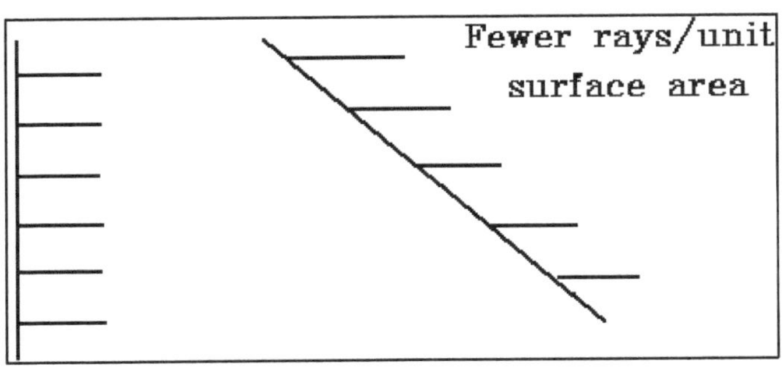

Define N = the unit normal to the surface and L = the unit vector which is the direction to the light source.

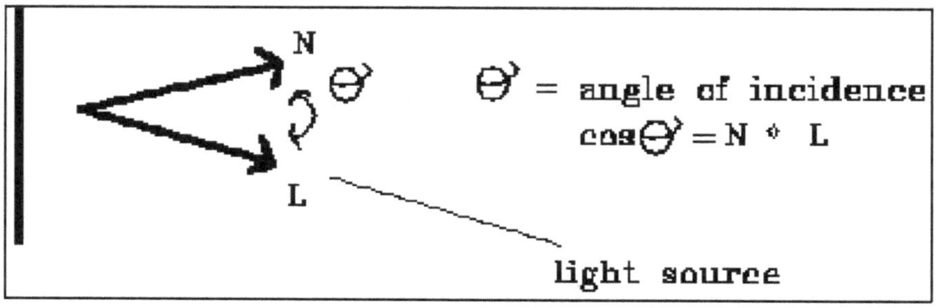

Then Lambert's law states that the reflected light is proportional to cos q (with q <= 90 else the light is behind the surface). Let Ip="intensity" of point light source, then the brightness is $f(1/d^2)$ where d="distance" of light source from the surface. Some models use a d distance dependence rather than d^2, and since most light sources are not really point sources, this works. Some models define a "distant" light source, such as the sun, which have no distance dependence.

Then I = [(kd * Ip) / (D)] * (N�L) with D = the distance function (< = 1 for a distant light source, d or d^2 for other models and kd = the diffuse coefficient of reflection with: 0.0 <= kd <= 1.0. For colored surfaces there are three kd values, one for red, green, and blue. Some models also add a constant term d_0 to prevent d the denominator from going to zero. Frequently ka is set to be the same as kd. Note that the vectors N and L are always normalized.

Then including ambient light we get a simple two term illumination model:

I = ka * Ia + [(kd * Ip) / (d)] * (N�L)

If we consider color then we should have an explicit wavelength dependence. But we usually make the approximation of sampling the color space at only three colors, i.e., "red", "green", and "blue", without really defining what wavelengths these are. Then we will have 3 equations like the above: one for Red, Green, and Blue. For example, a blue surface might have $kd^R = kd^G = 0$, $Kd^B = 0.8$. Note that the intensity, I, should always have a value between 0.0 and 1.0. It may be necessary to clamp the computed values so that they remain in the valid range.

Colored Objects

Up to this point white objects have been discussed, which do not absorb light. But the above scheme continues to be valid in the case that the material is absorbent. In this case, diffused rays will lose some wavelengths during their walk in the material, and will emerge colored.

Diffusion affects the color of objects in a substantial manner because it determines the average path of light in the material, and hence to which extent the various wavelengths are absorbed. Red ink looks black when it stays in its bottle. Its vivid color is only perceived when it is placed on a scattering material (e.g. paper). This is so because light's path through the paper fibers (and through the ink) is only a fraction of millimeter long. However, light from the bottle has crossed several centimeters of ink and has been heavily absorbed, even in its red wavelengths.

And, when a colored object has both diffuse and specular reflection, usually only the diffuse component is colored. A cherry reflects diffusely red light, absorbs all other colors and has a specular reflection which is essentially white (if the incident light is white light). This is quite general, because, except for metals, the reflectivity of most materials depends on their refractive index, which varies little with the wavelength (though it is this variation that causes the chromatic dispersion in a prism), so that all colors are reflected nearly with the same intensity. Reflections from different origin, instead, may be colored: metallic reflections, such as in gold or copper, or interferential reflections: iridescences, peacock feathers, butterfly wings, beetle elytra, or the anti-reflection coating of a lens.

Importance for Vision

Looking at one's surrounding environment, the vast majority of visible objects are seen primarily by diffuse reflection from their surface. This holds with few exceptions, such as glass, reflective liquids, polished or smooth metals, glossy objects, and objects that themselves emit light: the Sun, lamps, and computer screens (which, however, emit *diffuse* light). Outdoors it is the same, with perhaps the exception of a transparent water stream or of the iridescent colors of a beetle. Additionally, Rayleigh scattering is responsible for the blue color of the sky, and Mie scattering for the white color of the water droplets of clouds.

Light scattered from the surfaces of objects is by far the primary light which humans visually observe.

Interreflection

Diffuse interreflection is a process whereby light reflected from an object strikes other objects in the surrounding area, illuminating them. Diffuse interreflection specifically describes light reflected from objects which are not shiny or specular. In real life terms what this means is that light is reflected off non-shiny surfaces such as the ground, walls, or fabric, to reach areas not directly in view of a light source. If the diffuse surface is colored, the reflected light is also colored, resulting in similar coloration of surrounding objects.

In 3D computer graphics, diffuse interreflection is an important component of global illumination. There are a number of ways to model diffuse interreflection when rendering a scene. Radiosity and photon mapping are two commonly used methods.

Signal Reflection

Signal reflection is the process of bouncing a signal from a medium that does not totally absorb it. It may occur in copper cables for electrical signals, and in optical fiber for laser or optical signals. Signal reflection may also occur in outdoor metallic surfaces for electromagnetic (EM) waves. The EM waves travel through most open spaces and are not visible.

The study of signal reflection is used in specialized applications. Signal as sound may be reflected by a rigid surface and return to a receiver as a sound navigation and ranging (SONAR) signal. The ground-penetrating radar makes use of the principle that different radio frequencies and different ground materials will produce different amounts of signal absorption and reflection. In impedance matching, the goal is to make sure that most of the signal reaches the destination or load. The source impedance usually has to match the destination or load impedance for a given sub-band of radio frequencies.

In analog transmission cables, signal reflection is experienced as an echo when there is mismatch in the audio. Most of the issues in audio transmission have been resolved by using digitized audio in the form of Internet protocol data packets. Any signal reflection will be seen as data errors and eliminated with error correction schemes. Crosstalk, which used to be the unwanted induction of one analog signal from one cable to another, is also eliminated by using digital audio like the voice over Internet protocol (VoIP) packets in a digital subscriber line.

The Bergeron diagram shows the resulting voltages and currents when reflected electrical energy-combines with the incident energy. For best signal integrity, there should be minimal reflection, which is achieved by impedance matching. In some cases, the addition of resistive components that absorb electrical energy may totally eliminate reflection, while in other cases complex impedances formed by the series-parallel combinations of inductors and capacitors may provide the solution. The presence of distributed inductance and capacitance that is dependent on frequency makes the design of good impedance matching circuits very challenging.

Other specialized signal reflection methods include optical ranging where timed beam of light is allowed to reflect to a ranging target. Given the speed of light and the time it takes to receive the reflection, the distance to target can be computed. In radio detection and ranging (RADAR), the target reflects radio signals when the radar equipment sends a burst of radio frequency. The equipment waits for any reflected signal and computes the distance based on the delay between the transmission of radio frequency burst and reception of reflected signal, and the speed of radio waves in air.

Total Internal Reflection

Total internal reflection, in physics, complete reflection of a ray of light within a medium such as water or glass from the surrounding surfaces back into the medium. The phenomenon occurs if the angle of incidence is greater than a certain limiting angle, called the critical angle. In general, total internal reflection takes place at the boundary between two transparent media when a ray of light in a medium of higher index of refraction approaches the other medium at an angle of incidence greater than the critical angle. For a water-air surface the critical angle is 48.5°. Because indices of refraction depend on wavelength, the critical angle (and hence the angle of total internal reflection) will vary slightly with wavelength and, therefore, with colour. At all angles less than the critical angle, both refraction and reflection occur in varying proportions.

Glass prisms can be shaped to produce total internal reflection and as such are employed in binoculars, periscopes, telescopes, and other optical instruments. Light rays may be conducted over long, twisting paths by multiple total internal reflection in glass or plastic rods or fibres.

Two Requirements for Total Internal Reflection

Total internal reflection (TIR) is the phenomenon that involves the reflection of all the incident light off the boundary. TIR only takes place when both of the following two conditions are met:

- The light is in the more dense medium and approaching the less dense medium.
- The angle of incidence is greater than the so-called critical angle.

Total internal reflection will not take place unless the incident light is traveling within the more optically dense medium towards the less optically dense medium. TIR will happen for light traveling from water towards air, but it will not happen for light traveling from air towards water. TIR would happen for light traveling from water towards air, but it will not happen for light traveling from water (n=1.333) towards crown glass (n=1.52). TIR occurs because the angle of refraction reaches a 90-degree angle before the angle of incidence reaches a 90-degree angle. The only way for the angle of refraction to be greater than the angle of incidence is for light to bend away from the normal. Since light only bends away from the normal when passing from a more dense medium into a less dense medium, then this would be a necessary condition for total internal reflection.

Total internal reflection only occurs with large angles of incidence. Question: How large is large? Answer: larger than the critical angle. As mentioned above, the critical angle for the water-air boundary is 48.6 degrees. So for angles of incidence greater than 48.6-degrees, TIR occurs. But 48.6 degrees is the critical angle only for the water-air boundary. The actual value of the critical angle is dependent upon the two materials on either side of the boundary. For the crown glass-air boundary, the critical angle is 41.1 degrees. For the diamond-air boundary, the critical angle is 24.4 degrees. For the diamond-water boundary, the critical angle is 33.4 degrees. The critical angle is different for different media.

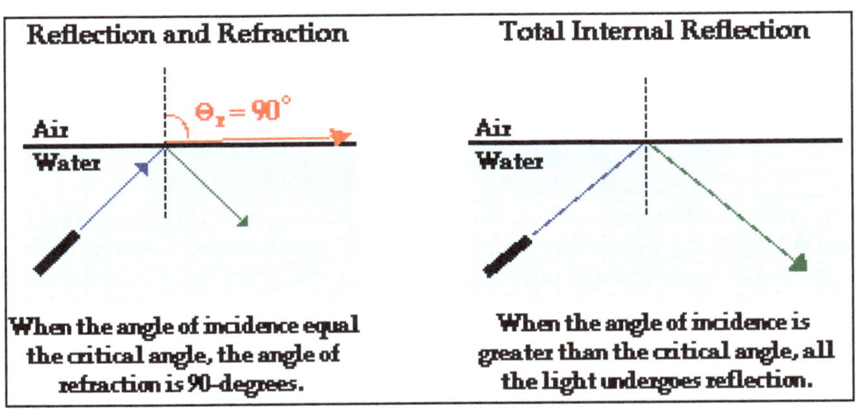

Optical Description

Total internal reflection of light can be demonstrated using a semi-circular block of glass or plastic. A "ray box" shines a narrow beam of light (a "ray") onto the glass medium. The semi-circular shape ensures that a ray pointing towards the centre of the flat face will hit the curved surface at a right angle; this will prevent refraction at the air/glass boundary of the curved surface. At the glass/air boundary of the flat surface, what happens will depend on the angle. If θ_c is the critical angle, then the following scenarios depict what will happen according to the size of the incident angle.

- If $\theta \le \theta_c$, the ray will split; some of the ray will reflect off the boundary, and some will refract as it passes through. This is not total internal reflection.
- If $\theta > \theta_c$, the entire ray reflects from the boundary. None passes through. This is called total internal reflection. TIR is the abbreviation.

This physical property makes optical fibers useful and prismatic binoculars possible. It is also what gives diamonds their distinctive sparkle, as diamond has an unusually high refractive index.

Critical Angle

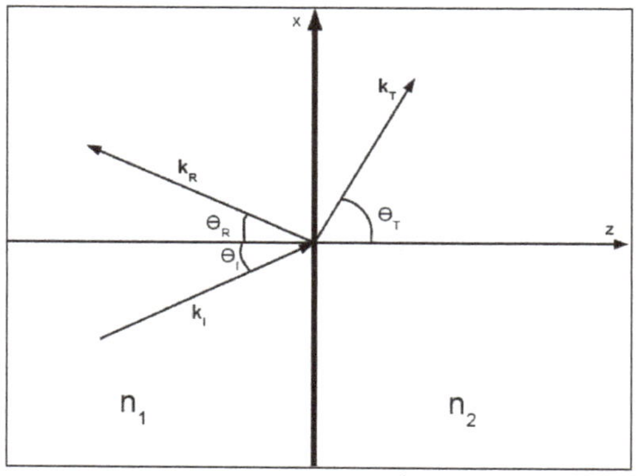

Illustration of Snell's law, $n_1 \sin\theta_i = n_2 \sin\theta_t$.

View up from directly under a diver. The above-water hemisphere is visible, compressed (as by a circular fisheye lens) into a circle (Snell's window) bounded by the critical angle. Everything outside the critical-angle circle is reflected from below the water (and is therefore dark).

The critical angle is the angle of incidence for which the angle of refraction is 90°. The angle of incidence is measured with respect to the normal at the refractive boundary. Consider a light ray passing from glass into air. The light emanating from the interface is bent towards the glass. When the incident angle is increased sufficiently, the transmitted angle (in air) reaches 90 degrees. It is at this point no light is transmitted into air. The critical angle θ_c is given by Snell's law,

$$n_1 \sin\theta_i = n_2 \sin\theta_t.$$

Rearranging Snell's Law, we get incidence,

$$\sin\theta_i = \frac{n_2}{n_1}\sin\theta_t.$$

To find the critical angle, we find the value for θ_i when $\theta_t = 90°$ and thus $\sin\theta_t = 1$. The resulting value of θ_i is equal to the critical angle θ_c.

Now, we can solve for θ_i, and we get the equation for the critical angle:

$$\theta_c = \theta_i = \arcsin\left(\frac{n_2}{n_1}\right),$$

If the incident ray is precisely at the critical angle, the refracted ray is tangent to the boundary at the point of incidence. If for example, visible light were traveling through acrylic glass(with an index of refraction of approximately 1.50) into air (with an index of refraction of 1.00), the calculation would give the critical angle for light from acrylic into air, which is,

$$\theta_c = \arcsin\left(\frac{1.00}{1.50}\right) = 41.8^\circ.$$

Light incident on the border with an angle less than 41.8° would be partially transmitted, while light incident on the border at larger angles with respect to normal would be totally internally reflected.

If the fraction n_2 / n_1 is greater than 1, then arcsine is not defined—meaning that total internal reflection does not occur even at very shallow or grazing incident angles.

So the critical angle is only defined when n_2 / n_1 is less than or equal to 1.

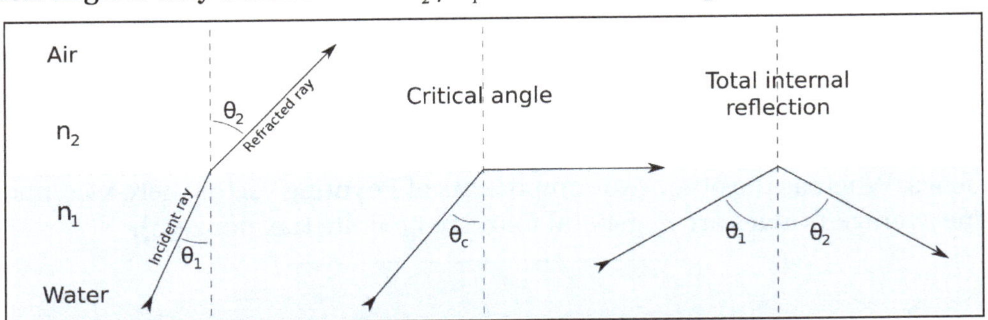

Refraction of light at the interface between two media, including total internal reflection.

Derivation of Evanescent Wave

An important side effect of total internal reflection is the appearance of an evanescent wave beyond the boundary surface. Essentially, even though the entire incident wave is reflected back into the originating medium, there is some penetration into the second medium at the boundary. The evanescent wave appears to travel along the boundary between the two materials, leading to the Goos-Hänchen shift.

If a plane wave, confined to the xz plane, is incident on a dielectric with an angle θ_I and wavevector $\mathbf{k}_\mathrm{I}$ then a transmitted ray will be created with a corresponding angle of transmittance as shown in the figure above. The transmitted wavevector is given by:

$$\mathbf{k}_\mathrm{T} = k_T \sin(\theta_T)\hat{x} + k_T \cos(\theta_T)\hat{z}$$

If $n_1 > n_2$, then $\sin(\theta_T) > 1$ since in the relation $\sin(\theta_T) = \frac{n_1}{n_2}\sin(\theta_I)$ obtained from Snell's law, $\frac{n_1}{n_2}\sin(\theta_I)$ is greater than one for $\theta_I > \theta_C$

As a result of this $\cos(\theta_T)$ becomes complex:

$$\cos(\theta_T) = \sqrt{1-\sin^2(\theta_T)} = \sqrt{\sin^2(\theta_T)-1}$$

The electric field of the transmitted plane wave is given by $\mathbf{E_T} = \mathbf{E_o}e^{i(\mathbf{k_T}\cdot\mathbf{r}-\omega t)}$ and so evaluating this further one obtains:

$$\mathbf{E_T} = \mathbf{E_o}e^{i(\mathbf{k_T}\cdot\mathbf{r}-\omega t)} = \mathbf{E_o}e^{i(xk_T \sin(\theta_T)+zk_T\cos(\theta_T)-\omega t)}$$

and ,

$$\mathbf{E_T} = \mathbf{E_o}e^{i(xk_T \sin(\theta_T)+zk_T i\sqrt{\sin^2(\theta_T)-1}-\omega t)}.$$

Using the fact that $k_T = \frac{\omega n_2}{c}$ and Snell's law, one finally obtains ,

$$\mathbf{E_T} = \mathbf{E_o}e^{-\kappa z}e^{i(kx-\omega t)}$$

where $\kappa = \frac{\omega}{c}\sqrt{(n_1\sin(\theta_I))^2 - n_2^2}\,a$ and $k = \frac{\omega n_1}{c}\sin(\theta_I)$.

This wave in the optically less dense medium is known as the evanescent wave. It is characterized by its propagation in the x direction and its exponential attenuation in the z direction. Although there is a field in the second medium, it can be shown that no energy flows across the boundary. The component of Poynting vector in the direction normal to the boundary is finite, but its time average vanishes. Whereas the other two components of Poynting vector (here x-component only) and their time averaged values are in general found to be finite (i.e. nonzero).

When a glass of water is held firmly, ridges making up the fingerprints are made visible by frustrated total internal reflection. Light tunnels from the glass into the ridges through the very short air gap.

It is important to note that the Goos-Hänchen effect only occurs for linearly polarized light. If the light is circularly or elliptically polarized, it will undergo the analogous Imbert–Fedorov effect.

Frustrated Total Internal Reflection

Under ordinary conditions, the evanescent wave transmits zero net energy across the interface. However, if a third medium with a higher refractive index than the low-index second medium is placed within less than several wavelengths distance from the interface between the first medium and the second medium, the evanescent wave will be different from the usual one, and it will pass energy across the second into the third medium. This process is called "frustrated" total internal reflection (FTIR) and is very similar to quantum tunneling. The quantum tunneling model is mathematically analogous if one thinks of the electromagnetic field as being the wave function of the photon. The low index medium can be thought of as a potential barrier through which photons can tunnel.

The transmission coefficient for FTIR is highly sensitive to the spacing between the third medium and the second medium (the function is approximately exponential until the gap is almost closed), so this effect has often been used to modulate optical transmission and reflection with a large dynamic range.

Phase Shift Upon Total Internal Reflection

An important aspect of total internal reflection is the appearance of a significant (rather than just 0° or 180°) phase shift upon reflection. Mathematically this is expressed by the Fresnel reflection coefficients acquiring an imaginary part. This resulting phase advance grows as the incidence angle increases beyond the critical angle, but unequally for the *s* and *p* polarization components of the incident wave.

This polarization dependent phase shift was discovered and quantitatively explained by Augustin-Jean Fresnel who employed it in his invention, the Fresnel rhomb, in which the angles of incidence are adjusted such that the two internal reflections cause a *relative* phase shift of 90° between the two polarizations of an incident wave. This performs the same function as a quarter-wave plate, but is more achromatic (phase shift has less dependence on wavelength). Either type of quarter-wave retarder may be used, for instance, to transform linear polarization to circular polarization (which Fresnel also discovered) and vice versa.

The polarization dependent phase shift is also the reason why TE and TM guided modes have different dispersion relations.

Applications

- Total internal reflection is the operating principle of optical fibers, which are used in endoscopes and telecommunications.
- Total internal reflection is the operating principle of automotive rain sensors, which control automatic windscreen/windshield wipers.
- Another application of total internal reflection is the spatial filtering of light.
- Prisms in binoculars use total internal reflection, rather than reflective coatings, to fold optical paths and show erect images.

- Some multi-touch screens use frustrated total internal reflection in combination with a camera and appropriate software to pick up multiple targets.
- Gonioscopy employs total internal reflection to view the anatomical angle formed between the eye's cornea and iris.
- A gait analysis instrument, CatWalk XT, uses frustrated total internal reflection in combination with a high speed camera to capture and analyze footprints of laboratory rodents.
- Optical fingerprinting devices use frustrated total internal reflection in order to record an image of a person's fingerprint without the use of ink.
- A total internal reflection fluorescence microscope uses the evanescent wave produced by TIR to excite fluorophores close to a surface. This is useful for the study of surface properties of biological samples.
- Total internal reflection is the operating principle of LED Light Panels. This technology utilizes LGPs (Light Guide Plates) as the vehicle for transmitting light over large areas. By etching a grid pattern on the second surface of the LGP, frustrated total internal reflection occurs allowing the light to escape the LGP as visible illumination.
- Diamonds shine brightly due to total internal reflection.
- Total internal reflection (TIR) are used on lighting industry to manage narrow distribution (NEMA classification) for floodlight.

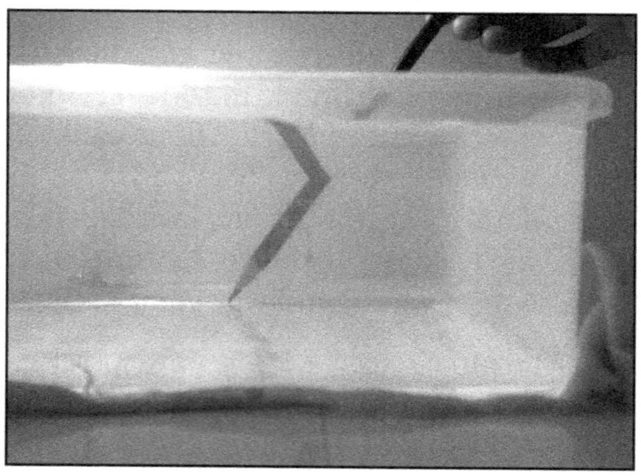

Mirror like effect.

Examples in Everyday Life

Total internal reflection can be observed while swimming, when one opens one's eyes just under the water's surface. If the water is calm, the surface appears mirror-like (except directly overhead, in Snell's window).

Diamonds used for jewelry are normally shaped in such a way as to maximize the amount of light that is totally internally reflected from the back side of the diamond. The very high refractive index of diamond gives a small critical angle allowing almost all the light entering the diamond to be reflected back for optimized diamond cuts such as the brilliant.

Total internal reflection can be seen at the air-water boundary.

Light Piping and Optical Fibers

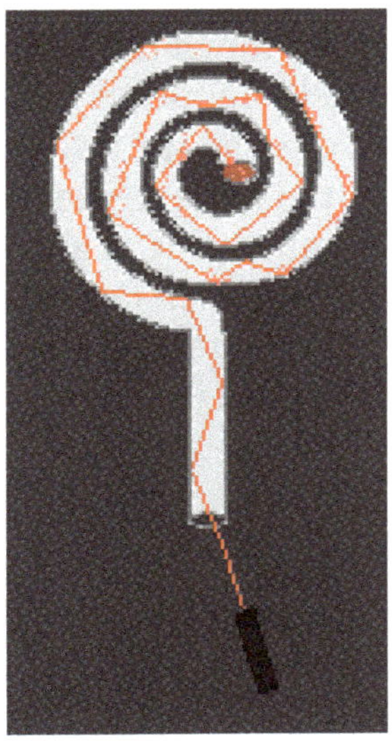

Total internal reflection is often demonstrated in a Physics class through a variety of demonstrations. In one such demonstration, a beam of laser light is directed into a coiled plastic *thing-a-ma jig*. The plastic served as a *light pipe*, directing the light through the coils until it finally exits out the opposite end. Once the light entered the plastic, it was in the more dense medium. Every time the light approached the plastic-air boundary, it is approaching at angles greater than the critical angle. The two conditions necessary for TIR are met, and all of the incident light at the plastic-air boundary stays internal to the plastic and undergoes reflection. And with the room lights off, every student becomes quickly aware of the ancient truth that Physics is better than drugs.

This demonstration helps to illustrate the principle by which optical fibers work. The use of a long strand of plastic (or other material such as glass) to pipe light from one end of the medium to the other is the basis for modern day use of optical fibers. Optical fibers are used in communication systems and micro-surgeries. Since total internal reflection takes place within the fibers, no incident energy is ever lost due to the transmission of light across the boundary. The intensity of the signal remains constant.

The laser beam stays internal to the water, continuously reflecting at each boundary.

Another common Physics demonstration involves the use of a large jug filled with water and a laser beam. The jug has a pea-sized hole drilled in its side such that when the cork is removed from the top of the jug, water begins to stream out the jug's side. The beam of laser light is then directed into the jug from the opposite side of the hole, through the water and into the falling stream. The laser light exits the jug through the hole but is still in the water. As the stream of water begins to fall as a projectile along a parabolic path to the ground, the laser light becomes trapped within the water due to total internal reflection. Being in the more dense medium (water) and heading towards a boundary with a less dense medium (air), and being at angles of incidence greater than the critical angle, the light never leaves the stream of water. In fact, the stream of water acts as a light pipe to pipe the laser beam along its trajectory. Once more, students viewing the demonstration are convinced of the fact that Physics is better than drugs.

Reflection Phase Change

Phase Change on Reflection

Idea: When light is reflected off a more dense medium with higher index of refraction, crests get reflected as troughs and troughs get reflected as crests. The wave is said undergo a 180 o change of phase on reflection. The net effect of the phase change is that the reflected ray "jumps ahead" by half a wave length. Thus, if one of the rays undergoes an odd number of phase changes, the conditions for constructive and destructive interference must be modified:

$$r_1 - r_2 = (m + \frac{1}{2})\lambda \;\; m = 1, 2, ... constructive$$

$$r_1 - r_2 = (m)\lambda \;\; m = 1, 2, ... destructive$$

- The above formulas are valid whenever an odd number of phase changes occur. If an even number of phase changes occur then the original, unmodified formulas must be used.
- The phase change does not happen when light is reflected off a less dense medium, i.e. for total internal reflection.

The phase changes that take place upon reflection play an important part in thin film interference.

Interference Condition for Thin Films

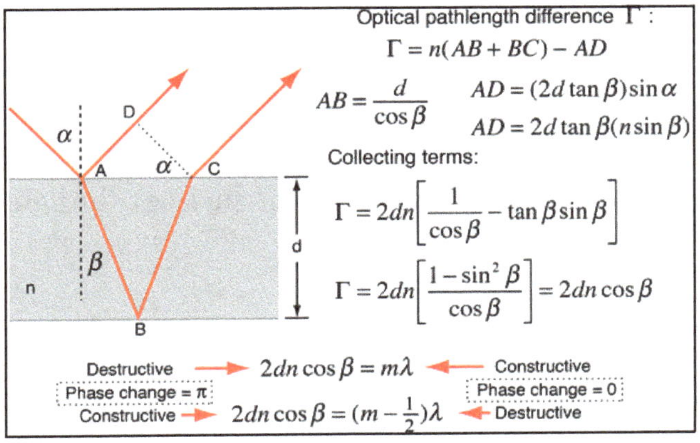

Refraction

Refraction is the bending of light (it also happens with sound, water and other waves) as it passes from one transparent substance into another.

This bending by refraction makes it possible for us to have lenses, magnifying glasses, prisms and rainbows. Even our eyes depend upon this bending of light. Without refraction, we wouldn't be able to focus light onto our retina.

Change of Speed Causes Change of Direction

Light refracts whenever it travels at an angle into a substance with a different refractive index (optical density).

This change of direction is caused by a change in speed. For example, when light travels from air into water, it slows down, causing it to continue to travel at a different angle or direction.

How much does light bend?

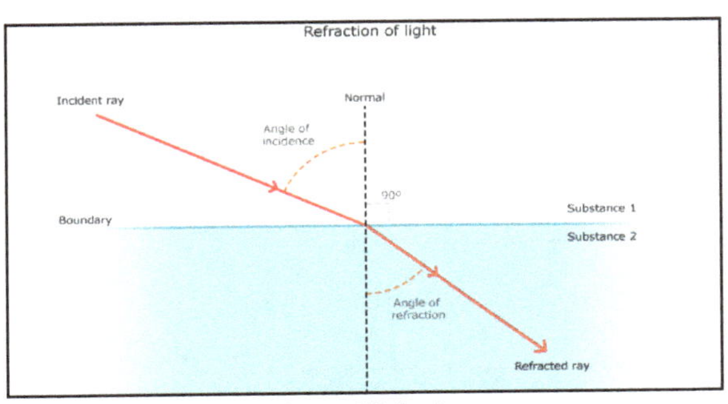

Refraction of light in water.

When light travels from air into water, it slows down, causing it to change direction slightly. This change of direction is called refraction. When light enters a more dense substance (higher refractive index), it 'bends' more towards the normal line.

The amount of bending depends on two things:

- Change in speed – if a substance causes the light to speed up or slow down more, it will refract (bend) more.
- Angle of the incident ray – if the light is entering the substance at a greater angle, the amount of refraction will also be more noticeable. On the other hand, if the light is entering the new substance from straight on (at 90° to the surface), the light will still slow down, but it won't change direction at all.

Refractive index of some transparent substances

Substance	Refractive index	Speed of light in substance (x 1,000,000 m/s)	Angle of refraction if incident ray enters substance at 20°
Air	1.00	300	20
Water	1.33	226	14.9
Glass	1.5	200	13.2
Diamond	2.4	125	8.2

All angles are measured from an imaginary line drawn at 90° to the surface of the two substances. This line is drawn as a dotted line and is called the normal.

If light enters any substance with a *higher* refractive index (such as from air into glass) it slows down. The light bends *towards* the normal line.

If light travels enters into a substance with a *lower* refractive index (such as from water into air) it speeds up. The light bends *away* from the normal line.

A higher refractive index shows that light will slow down and change direction more as it enters the substance.

Lenses

A lens is simply a curved block of glass or plastic. There are two kinds of lens.

A biconvex lens is thicker at the middle than it is at the edges. This is the kind of lens used for a magnifying glass. Parallel rays of light can be focused in to a focal point. A biconvex lens is called a converging lens.

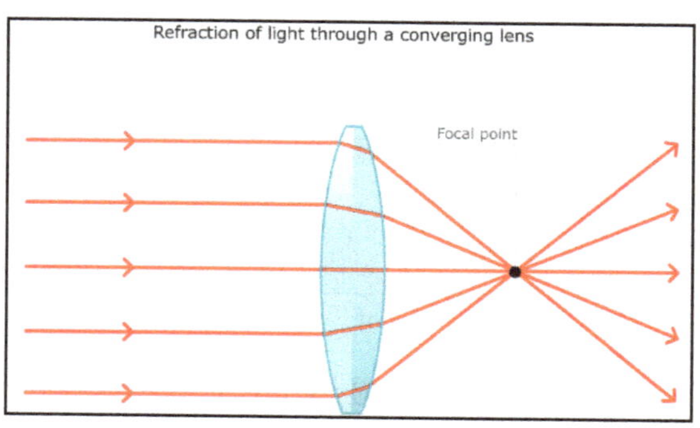

Each light ray entering a converging (convex) lens refracts inwards as it enters the lens and inwards again as it leaves. These refractions cause parallel light rays to spread out, travelling directly away from an imaginary focal point.

A biconcave lens curves is thinner at the middle than it is at the edges. Light rays refract outwards (spread apart) as they enter the lens and again as they leave.

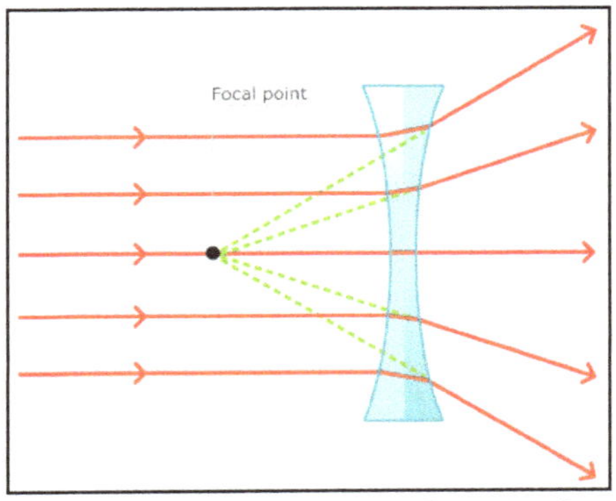

Refraction of light through a diverging lens.

Each light ray entering a diverging (concave) lens refracts outwards as it enters the lens and outwards again as it leaves. These refractions cause parallel light rays to spread out, travelling directly away from an imaginary focal point.

Refraction Can Create a Spectrum

Isaac Newton performed a famous experiment using a triangular block of glass called a prism. He used sunlight shining in through his window to create a spectrum of colours on the opposite side of his room.

When white light shines through a prism, each colour refracts at a slightly different angle. Violet light refracts slightly more than red light. A prism can be used to show the seven colours of the spectrum that make up white light.

This experiment showed that white light is actually made of all the colours of the rainbow. These seven colours are remembered by the acronym ROY G BIV – red, orange, yellow, green, blue, indigo and violet.

Newton showed that each of these colours cannot be turned into other colours. He also showed that they can be recombined to make white light again.

The explanation for the colours separating out is that the light is made of waves. Red light has a longer wavelength than violet light. The refractive index for red light in glass is slightly different than for violet light. Violet light slows down even more than red light, so it is refracted at a slightly greater angle.

The refractive index of red light in glass is 1.513. The refractive index of violet light is 1.532. This slight difference is enough for the shorter wavelengths of light to be refracted more.

Rainbows

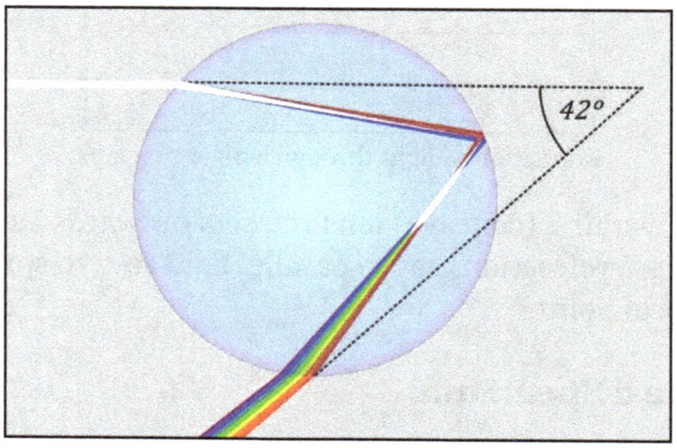

A rainbow is formed when light enters each water droplet, and the different colours bend (refract) at slightly different angles. They reflect off the inside of the drop before refracting again as they leave. The shorter wavelengths refract more.

A rainbow is caused because each colour refracts at slightly different angles as it enters, reflects off the inside and then leaves each tiny drop of rain.

A rainbow is easy to create using a spray bottle and the sunshine. The centre of the circle of the rainbow will always be the shadow of your head on the ground.

The secondary rainbow that can sometimes be seen is caused by each ray of light reflecting twice on the inside of each droplet before it leaves. This second reflection causes the colours on the secondary rainbow to be reversed. Red is at the top for the primary rainbow, but in the secondary rainbow, red is at the bottom.

Refractive Index

The refractive index or index of refraction (n_D) is one of the physicochemical properties of substances (optical medium). It is equal to velocity (c) of light of a given wavelength in empty space (vacuum) divided by its velocity (v) in a substance according to following equation:

$$n_D = \frac{c}{v}$$

When light travels at an angle between two materials, it bends according to their refractive indices. In order to reflect, light must be on the wider side of the critical angle.

The refractive index can provide information for us about the behavior of light. When light passes through the different substances its velocity decreases by increasing of the refractive index of these substances. It can be due to interaction between molecules of components in substrate and effect of these interactions on light. Also in the most substrates, the refractive index decreases by increasing of the temperature. The interaction between molecules decreases as the temperature increases.

The refractive index of different substrates measures with refractometers. There are four main types of refractometers: traditional handheld refractometers, digital handheld refractometers, laboratory or Abbe refractometers, and inline process refractometers. There is also the Rayleigh Refractometer used (typically) for measuring the refractive indices of gases.A sodium lamp may be used to provide the light source at a known wave-length (589.6 nm) although many instruments are corrected for daylight use.

In most of substrates which we know the refractive index have a positive value (more than zero) but in 1968 Veselago shows that there are substrates with negative permittivity and negative permeability. In these substrates refractive index have a negative value.

Maxwell's equations relate the permittivity and the permeability to the refractive index as follows:

$$n = \pm\sqrt{s\mu}$$

The sign of the index is often taken as positive. However, Veselago showed that if a substrate has both negative permeability and negative permittivity, this convention must be reversed and we must pick out the *negative* sign of the square root.

The Speed of Light

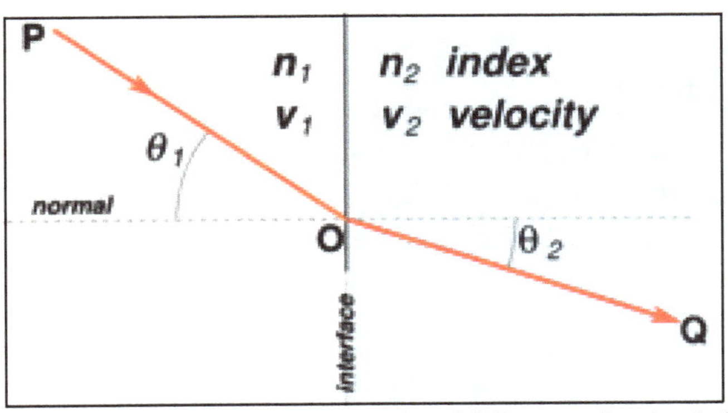

Refraction of light at the interface between two media of different refractive indices, with $n_2 > n_1$. The velocity is lower in the second medium ($v_2 < v_1$), therefore the angle of refraction θ_2 is less than the angle of incidence θ_1; that is, the ray in the higher-index medium is closer to the normal.

The speed of all electromagnetic radiation in vacuum is the same, approximately 3×10^8 meters per second, and is denoted by *c*. Therefore, if *υ* is the phase velocity of radiation of a specific frequency in a specific material, the refractive index is given by,

$$n = \frac{c}{v} \quad .$$

This number is typically greater than one: the higher the index of the material, the more the light is slowed down. However, at certain frequencies (e.g., X-rays), n will actually be smaller than one. This does not contradict the theory of relativity, which holds that no information-carrying signal can ever propagate faster than c, because the phase velocity is not the same as the group velocity or the signal velocity, same as group velocity except when the wave is passing through an absorptive medium.

Sometimes, a "group velocity refractive index," usually called the *group index* is defined:

$$n_g = \frac{c}{v_g}$$

where v_g is the group velocity. This value should not be confused with n, which is always defined with respect to the phase velocity.

At the microscale, an electromagnetic wave's phase velocity is slowed in a material because the electric field creates a disturbance in the charges of each atom (primarily the electrons) proportional (a $y = kx$ relationship) to the permittivity. The charges will, in general, oscillate slightly out of phase with respect to the driving electric field. The charges thus radiate their own electromagnetic wave that is at the same frequency but with a phase delay. The macroscopic sum of all such contributions in the material is a wave with the same frequency but shorter wavelength than the original, leading to a slowing of the wave's phase velocity. Most of the radiation from oscillating material charges will modify the incoming wave, changing its velocity. However, some net energy will be radiated in other directions. If the refractive indices of two materials are known for a given frequency, then one can compute the angle by which radiation of that frequency will be refracted as it moves from the first into the second material from Snell's law.

Negative Refractive Index

Recent research has also demonstrated the existence of negative refractive index, which can occur if ε and μ are *simultaneously* negative. Not thought to occur naturally, it can be achieved with so called metamaterials. It offers the possibility of perfect lenses and other exotic phenomena such as a reversal of Snell's law.

References

- Geometrical-optics, reflection-at-smooth-surfaces: brilliant.org, Retrieved 07, February 2020
- Geometrical-optics: brilliant.org, Retrieved 11, May 2020
- Axelrod, D. "Cell-substrate contacts illuminated by total internal reflection fluorescence". The Journal of Cell Biology. 89 (1): 141–145. doi:10.1083/jcb.89.1.141. PMC 2111781 . PMID 7014571
- What-is-diffuse-reflection: wisegeek.com, Retrieved 15, March 2020

- Angle-of-incidence: light: physics.tutorvista.com, Retrieved 27, April 2020
- OD, By Emily Bruce, OD, Rodney Bendure, OD, Sarah Krein, OD, and Nathan Lighthizer,. "Zoom in on Gonioscopy". Retrieved 13, August 2020
- Law-of-reflection, light: physics.tutorvista.com, Retrieved 17, July 2020

Optical Devices

There are various applications of optics such as microscopes, optical telescopes, optical periscopes, cameras and optical computing, etc. Optics also has applications in the field of healthcare and life sciences, astronomy and space and industrial manufacturing. The diverse applications of optics in the current scenario have been thoroughly discussed in this chapter.

Optical Fiber

Optical fiber is the technology associated with data transmission using light pulses travelling along with a long fiber which is usually made of plastic or glass. Metal wires are preferred for transmission in optical fiber communication as signals travel with fewer damages. Optical fibers are also unaffected by electromagnetic interference. The fiber optical cable uses the application of total internal reflection of light. The fibers are designed such that they facilitate the propagation of light along with the optical fiber depending on the requirement of power and distance of transmission. Single-mode fiber is used for long-distance transmission, while multimode fiber is used for shorter distances. The outer cladding of these fibers needs better protection than metal wires.

Construction

An optical fiber consists of three basic concentric elements: the core, the cladding, and the outer coating. The core is usually made of glass or plastic, although other materials are sometimes used, depending on the transmission spectrum desired.

The core is the light-transmitting portion of the fiber. The cladding usually is made of the same material as the core, but with a slightly lower index of refraction (usually about 1% lower). This index difference causes total internal reflection to occur at the index boundary along the length of the fiber so that the light is transmitted down the fiber and does not escape through the sidewalls.

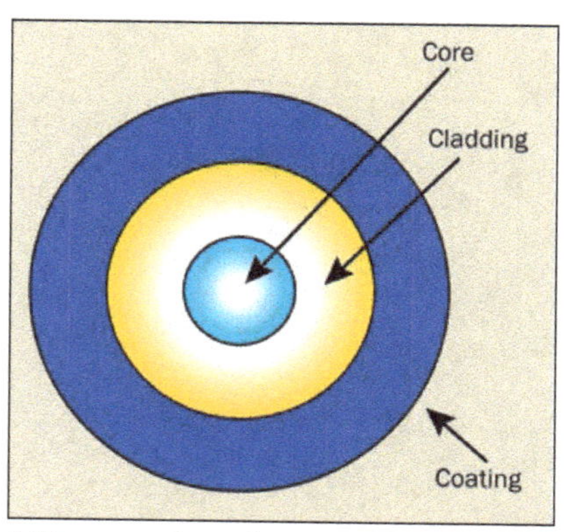

An optical fiber consists of a core, cladding, and coating.

The coating usually comprises one or more coats of a plastic material to protect the fiber from the physical environment. Sometimes metallic sheaths are added to the coating for further physical protection. Optical fibers usually are specified by their size, given as the outer diameter of the core, cladding, and coating. For example, a 62.5/125/250 would refer to a fiber with a 62.5-µm diam core, a 125-µm diam cladding, and a 0.25-mm diam outer coating.

Principles

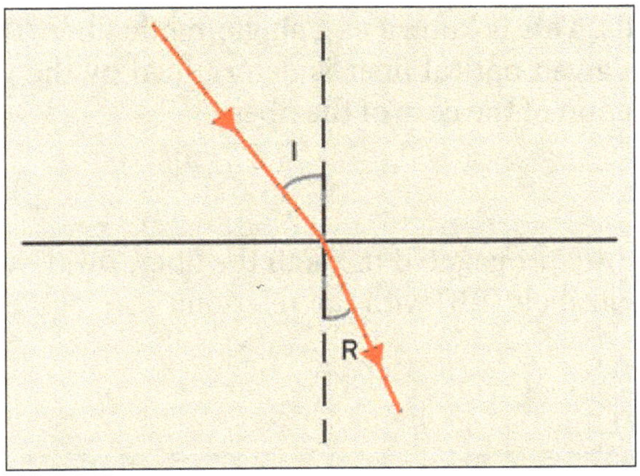

A beam of light passing from one material to another of a different index of refraction is bent or refracted at the interface.

Optical materials are characterized by their index of refraction, referred to as n. A material's index of refraction is the ratio of the speed of light in a vacuum to the speed of light in the material. When a beam of light passes from one material to another with a different index of refraction, the beam is bent (or refracted) at the interface.

Refraction is described by Snell's law:

$$n_I \sin I = n_R \sin R$$

where n_I and n_R are the indices of refraction of the materials through which the beam is refracted and I and R are the angles of incidence and refraction of the beam. If the angle of incidence is greater than the critical angle for the interface (typically about 82° for optical fibers), the light is reflected back into the incident medium without loss by a process known as total internal reflection.

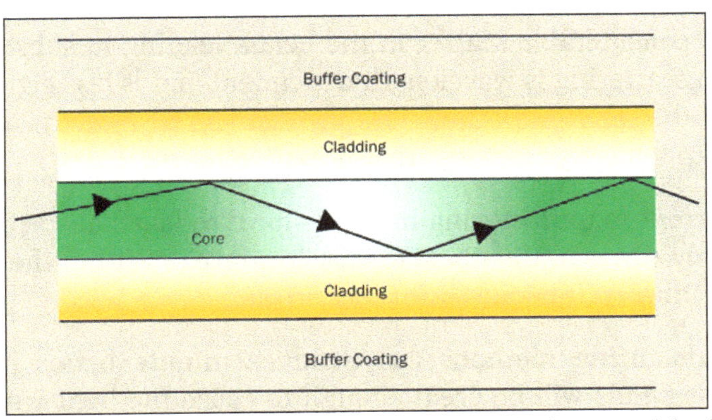

Total internal reflection allows light to remain inside the core of the fiber.

Modes

When light is guided down a fiber (as microwaves are guided down a waveguide), phase shifts occur at every reflective boundary. There is a finite discrete number of paths down the optical fiber (known as modes) that produce constructive (in phase and therefore additive) phase shifts that reinforce the transmission. Because each mode occurs at a different angle to the fiber axis as the beam travels along the length, each one travels a different length through the fiber from the input to the output. Only one mode, the zero-order mode, travels the length of the fiber without reflections from the sidewalls. This is known as a single-mode fiber. The actual number of modes that can be propagated in a given optical fiber is determined by the wavelength of light and the diameter and index of refraction of the core of the fiber.

Attenuation

Signals lose strength as they are propagated through the fiber; this is known as beam attenuation. Attenuation is measured in decibels (dB) with the relation:

$$A(dB) = 10 \log\left(\frac{P_{in}}{P_{out}}\right)$$

where P_{in} and P_{out} refer to the optical power going into and coming out of the fiber. The table below shows the power typically lost in a fiber for several values of attenuation in decibels. The attenuation of an optical fiber is wavelength dependent. At the extremes of the transmission curve, multiphoton absorption predominates. Attenuation is usually expressed in dB/km at a specific wavelength. Typical values range from 10 dB/km for step-index fibers at 850 nm to a few tenths of a dB/km for single-mode fibers at 1550 nm.

Attenuation (dB)	Power Loss (%)
10.0	90
3.0	50
0.1	2
Power loss as a function of attenuation	

There are several causes of attenuation in an optical fiber:

- Rayleigh Scattering: Microscopic-scale variations in the index of refraction of the core material can cause considerable scatter in the beam, leading to substantial losses of optical power. Rayleigh scattering is wavelength dependent and is less significant at longer wavelengths. This is the most important loss mechanism in modern optical fibers, generally accounting for up to 90% of any loss that is experienced.

- Absorption: Current manufacturing methods have reduced absorption caused by impurities (most notably water in the fiber) to very low levels. Within the bandpass of transmission of the fiber, absorption losses are insignificant.

- Bending: Manufacturing methods can produce minute bends in the fiber geometry. Sometimes these bends will be great enough to cause the light within the core to hit the core/cladding interface at less than the critical angle so that light is lost into the cladding material. This also can occur when the fiber is bent in a tight radius (less than, say, a few centimeters). Bend sensitivity is usually expressed in terms of dB/km loss for a particular bend radius and wavelength.

Numerical Aperture

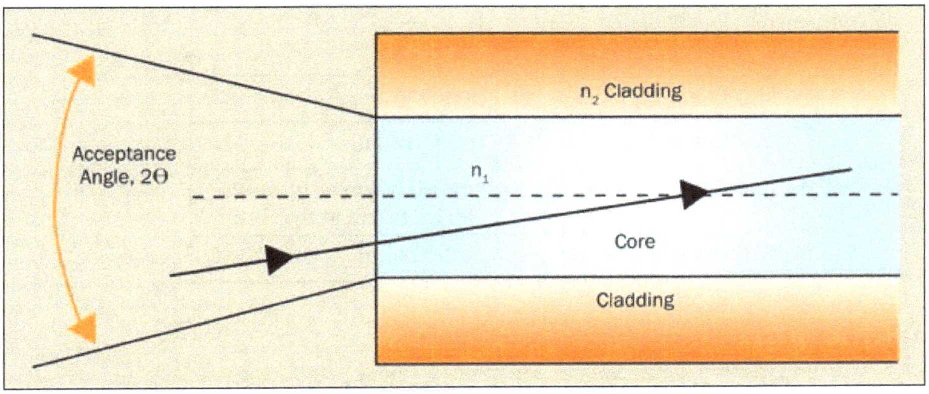

Numerical aperture depends on the angle at which rays enter the fiber and on the diameter of the fiber's core.

Numerical aperture (NA) is the measure of maximum angle at which light rays will enter and be conducted down the fiber. This is represented by the following equation:

$$NA = \sqrt{\left(n_{core}^2 - n_{cladding}^2\right)} = \sin\theta$$

Dispersion

As the optical pulses travel the length of the fiber, they are broadened or lengthened in time. This is called dispersion. Because the pulses eventually will become so out of step that they begin to overlap each other and corrupt the data, dispersion sets an upper limit on the data-carrying capabilities of a fiber. There are three principal causes for this broadening:

- Chromatic Dispersion: Different wavelengths travel at different velocities down the fiber. Because typical light sources provide power over a series or range of wavelengths, rather than from a single discrete spectral line, the pulses must spread out along the length of the fiber as they proceed. The high-speed lasers used in communications have very narrow spectral output specifications, greatly reducing the effect of chromatic dispersion.
- Modal Dispersion: Different fiber modes reflect at different angles as they proceed down the fiber. Because each modal angle produces a somewhat different path length for the beam, the higher-order modes reach the output end of the fiber behind the lower-order modes.
- Waveguide Dispersion: This minor cause for dispersion is due to the geometry of the fiber and results in different propagation velocities for each of the modes.

Bandwidth

Bandwidth measures the data-carrying capacity of an optical fiber and is expressed as the product of the data frequency and the distance traveled (MHz-km or GHz-km, typically). For example, a fiber with a 400-MHz-km bandwidth can transmit 400 MHz for a distance of 1 km, or it can transmit 20 MHz of data for 20 km. The primary limit on bandwidth is pulse broadening, which results from modal and chromatic dispersion of the fiber. Typical values for different types of fiber follow:

Fiber Type	Bandwidth
Single Mode	100 GHz-km
Graded Index	500 MHz-km at 1300 nm 160 MHz-km at 850 nm
Step Index	20 MHz-km

Power Transmission

The amount of power that a fiber can transmit (without being damaged) is usually expressed in terms of the maximum acceptable power density. Power density is the product of the maximum power output of the laser and the area of the laser beam. For example, a 15-W laser beam focused onto a 150-µm diameter spot produces a power density of,

$$\frac{P}{A} = \frac{15\,W}{\pi(0.0075\,cm)^2} = 85\,kW/cm^2$$

The output of a pulsed laser (typically specified in millijoules of energy per pulse) must first be converted to power per pulse. For example, a pulsed laser that produces 50 mJ in a 10-ns pulse provides an output power of,

$$P = \frac{50 \times 10^{-3}\,J}{10^{-8}\,s} = 5000\,kW = 5\,MW$$

The power density then can be calculated from the spot size. To transmit the absolute maximum energy levels down a fiber, the fiber end faces must be absolutely smooth and polished and be perpendicular to the fiber axis and the light beam. Also, the beam diameter should be no greater than approximately one-half of the area of the core (or the diameter of the core). If the beam is not appropriately focused, some of the energy may spill into the cladding, which quickly can damage polymer-clad silica fibers. For this reason, it is better to use silica-clad silica fibers in higher power density applications.

Fiber Types

There are basically three types of optical fiber: Single-mode, multimode graded index, and multimode step-index. They are characterized by the way light travels down the fiber and depend on both the wavelength of the light and the mechanical geometry of the fiber. Examples of how they propagate light are shown in figure.

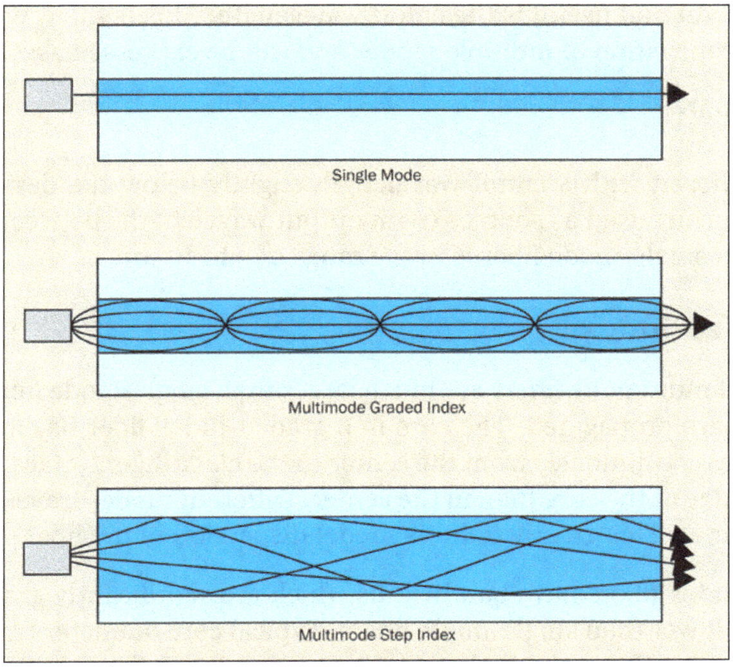

Modes of fiber transmission.

Single-mode

Only the fundamental zero-order mode is transmitted in a single-mode fiber. The light beam travels straight through the fiber with no reflections from the core-cladding sidewalls at all. Single-mode fiber is characterized by the wavelength cutoff value, which is dependent on core diameter, NA and wavelength of operation. Below the cutoff wavelength, higher-order modes may also propagate, which changes the fiber's characteristics.

Because the single-mode fiber propagates only the fundamental mode, modal dispersion (the primary cause of pulse overlap) is eliminated. Thus, the bandwidth is much higher with a single-mode fiber than that of a multimode fiber. This simply means that pulses can be transmitted much closer together in time without overlap. Because of this higher bandwidth, single-mode fibers are used in all modern long-range communication systems. Typical core diameters are between 5 and 10 µm.

The actual number of modes that can be propagated through a fiber depends on the core diameter, the numerical aperture and the wavelength of the light being transmitted. These may be combined into the normalized frequency parameter or V number,

$$V = \frac{2\pi a}{\lambda}\sqrt{\left(n_{core}^2 - n_{cladding}^2\right)} = \frac{2\pi a}{\lambda}NA$$

where a is the core radius, λ is the wavelength, and n is the index of the core and the cladding. The condition for single-mode operation is that:

$$V \le V_{cutoff} = 2.405$$

Perhaps more important and useful is the cutoff wavelength. This is the wavelength below which the fiber will allow propagation of multiple modes and can be expressed as:

$$\lambda \geq \lambda_{cutoff} = 2.6\,aNA$$

A fiber is typically chosen with a cutoff wavelength slightly below the desired operating wavelength. For lasers typically used as sources (with output wavelengths between 850 and 1550 nm), the core diameter of a single-mode fiber is in the range of 3 to 10 μm.

Multimode Graded-index

The core diameters of multimode fibers are much larger than single-mode fibers. As a result, higher-order modes also are propagated. The core in a graded-index fiber has an index of refraction that radially decreases continuously from the center to the cladding interface. As a result, the light travels faster at the edge of the core than in the center. Different modes travel in curved paths with nearly equal travel times. This greatly reduces modal dispersion in the fiber.

As a result, graded-index fibers have bandwidths which are significantly greater than step-index fibers, but still much lower than single-mode fibers. Typical core diameters of graded-index fibers are 50, 62.5, and 100 μm. The main application for graded-index fibers is in medium-range communications, such as local area networks.

Multimode Step-index

The core of a step-index fiber has a uniform index of refraction right up to the cladding interface where the index changes in a step-like fashion. Because different modes in a step-index fiber travel different path lengths in their journey through the fiber, data transmission distances must be kept short to avoid considerable modal dispersion problems. Step-index fibers are available with core diameters of 100 to 1500 μm. They are well suited to applications requiring high power densities, such as medical and industrial laser power delivery.

Importance of Fiber Optic Cables

Before we move on to the importance of fiber cables, it is best to first understand the various industries and uses of fiber optic cables. The cabling serves the following purposes:

- Internet connection, which connects everyone (more quickly) at affordable costs.
- Connecting telephones and modern-age telephony.
- Connecting computer networks together to boost the speed of communication.
- Used in the medical fields of dentistry and surgery.
- Decorations and lighting when referencing household and general purpose illumination.
- Inspection of remote areas.
- Improving the quality of cable transmissions.

- The automotive industry, by providing lighting and safety features.
- Military and space travel research, where transmission of data is fast and reliable within high-security installations.

Looking at the various uses, many might still be in doubt about the positive impact that fiber optics are playing in the other sectors, besides home and networking uses. The doubt might be due to the fact that they are not aware of what makes a fiber optic cable important, and what makes it stand out. The following are some considerations that come in to play when boosting the use of fiber optic cables:

- Faster Internet Speeds: When it comes to completing jobs in the workplace, internet speed is a must. Even at home, faster internet speeds carry much importance. The faster you are finished gathering research materials for your assignment, the quicker you will be able to go out and relax. The fiber optic provides a state of the art system of internet connectivity. The current lifestyle is very busy, and you need faster internet connection speeds that promote convenience and reliability.
- Have Wider Coverage: This is in reference to the amount of distance that your internet connection covers. Fiber optics carry signals to greater distances in a shorter timespan compared to other cabling options. Though there is a slight difference in the distance they cover, for factors like cable type, network option, and wavelengths, the distance is significantly longer. In this, the mode of communication is eased since the coverage is reduced, giving you the ability to reach out to more individuals and expand your network. This is not only significant in a business setting, but it is also very useful when you are seeking to video call your relatives.
- Greater and better bandwidth: A better bandwidth is crucial since it provides you with the ability to transmit more data when considering a similar diameter. With fiber optics, the level of network connectivity between computers has also been made a lot easier and more convenient. You will not experience data congestion (waiting for hours) for complete transmission. Similarly, file transfers will also take a lot less time, so you can transfer lots of data as well.
- Lower Cost of Ownership: As every new piece of technology, fiber optic is not cheap. The cost of installation is relatively higher than other cables. However, when considering the overall cost of ownership, like installation, maintenance, and costs of use, you will find that it is cost effective. With maintenance, you will find that fiber optics have a lesser maintenance cost, making it durable in the long run. With the advancement of technology, maintenance costs are also reduced. Keep this in mind as well.
- Flexibility and Versatility: The current technological landscape is very quick in evolution, so users are looking for a network that keeps up with those changes. That being said, you can always upgrade your equipment to sync them with fiber optic cables.
- Reliability: Another benefit that you can gain from fiber optic cables is their reliability, which is resistant to changes in weather and temperature. The network connectivity is not interrupted due to moisture, heat, or even electromagnetic interference since fiber optic

cables do not transmit electric current. With this, you will not experience an interruption in data transmission, which is essential for professional use.

Microscopes

Microscopy studies the enlargement of the image of objects too small to be properly seen by the unaided eye. Microscopy accomplishes its task making use of the radiations emitted, absorbed, transmitted, or reflected by the specimen to be observed. The nature of the radiation specifies the type of microscopy: electron microscopy, x-ray microscopy, acoustic microscopy, etc. The visible part of electromagnetic spectrum is the type of radiation used by optical microscopy.

Thin Lenses

The foundation of the magnifying power of lenses is the refraction phenomenon. A light beam crossing the surface between two media of different optical densities is bent according to the law discovered by Willebrod Snell:

$$n_1 \sin \theta_1 = n_2 \sin \theta_2$$

where θ_1 and θ_2 are the angles formed by the ray and the normal to the surface at the point of incidence into mediums 1 and 2, respectively, and n_1 and n_2 are the refractive indexes. The refractive index n of a material is related to the velocity of light v in the same material: $v = c/n$, where c is light speed in the vacuum. The indices of refraction of some common substances are given below:

- Vacuum: 1.000
- Air: 1.0003
- Ice: 1.31
- Water: 1.33
- Ethyl alcohol: 1.36
- Glycerin: 1.47
- Crown glass: 1.50 ÷ 1.62
- Flint glass: 1.57 ÷ 1.75
- Polystyrene: 1.59
- Carbon disulfide: 1.63
- Methylene iodide: 1.74
- Diamond: 2.42

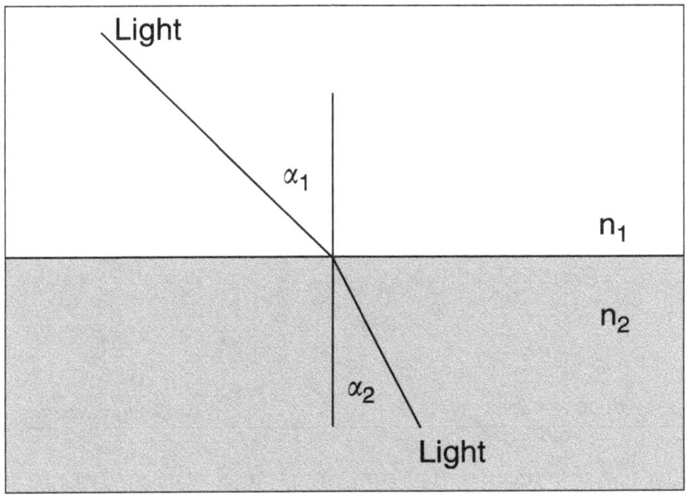

Light crossing the interface separating two transparent media of refractive index n_1 and n_2, respectively ($n_2 > n_1$ 9). The angle α_1 between the light ray and the normal to the interface is called angle of incidence. The angle α_2 is called angle of refraction.

According to this effect, the light crossing a thin lens passes from air into glass and then from glass into air, being refracted twice. Rays entering a lens parallel to the optical axis emerge from it along a path intersecting the focal point on the opposite side. Thin lenses have a focal point at every side, and the focal length f is the distance between the focal points and the lens. If o and i are the distances from the lens of object and image respectively, then $1/o + 1/i = 1/f$, and the magnification m is i/o. A thin lens divides the space in an incident- and in a refracted-light emispace. The values of o, i, m, and f can be positive (+) or negative (-) according to the following conventions:

- Incident-light emispace: o(+), i(-),
- Refracted-light emispace: o(-), i(+),
- Converging lenses: f(+),
- Diverging lenses: f(-),
- Image upright: m(+),
- Image inverted: m(-).

Images in the refracted-light emispace are real images; images in the incident-light emispace are virtual images. In microscopes, more than two refracting surfaces exist, because the simplest type of microscope is formed by two thin lenses, objective and eyepiece, where the image of the first is the object of the second one. In modern microscopes, both objectives and eyepieces are formed by many different groups of lenses; by assembling lenses in the right way, very high magnification values may be obtained. Moreover, modern objectives are infinity corrected. Light emerges from such objectives in parallel rays that are focalized by a tube lens at the intermediate image plane. This arrangement frees the manufacturers from any constraint in the optical tube length.

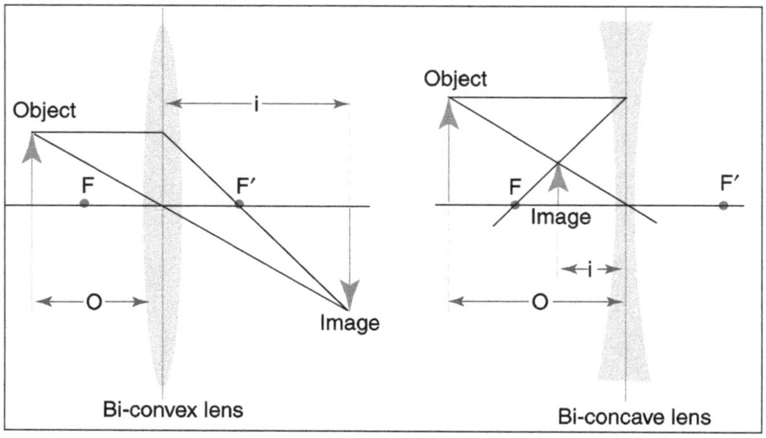

Image formation by thin lenses according to geometrical optics: Image is located according to three rules: (1) any incident ray traveling parallel to the principal axis is refracted through the focal point F in converging lenses and in a direction such that its extension passes through the focal point in diverging lenses; (2) light rays traveling through the center of the lens proceed in a straight line; and (3) any incident ray traveling through the focal point on the way to the lens is refracted parallel to the principal axis. According to conventions, the image on the left (converging lens) is real and inverted and the image on the right (diverging lens) is virtual and upright.

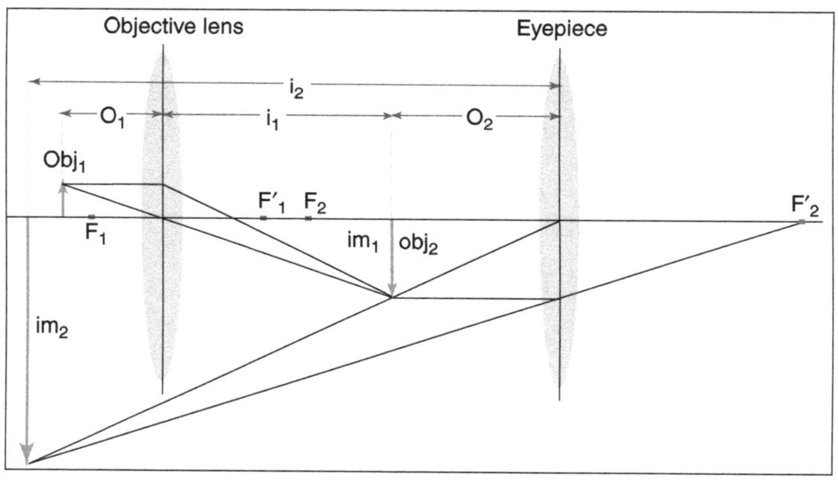

The simple microscope.

A microscope uses a very short focal length objective lens to form a greatly enlarged image. This image is then viewed with a short focal length eyepiece used as a simple magnifier. According to this assumption, the resulting magnification M is $M = m_1 \times m_2 = i_1/o_1 \times i_2/o_2 = i_1 \times i_2/o_1 \times o_2$. In modern microscopes, the image is formed at infinity to minimize eyestrain.

Optical Resolution

The ability of an optical instrument to produce separable images of different points on an object is limited. The resolving power of a lens is a quantitative measure of this ability: points closer than the limit of resolution cannot be distinguished as separate points. In 1873 first fixed the value of the minimal distance d between two adjacent points allowing them to be perceived as separated:

$$d = \lambda / 2n \sin \alpha,$$

where λ is the wavelength of light, α is one-half the angular aperture of the lens, and n is the refractive index of the medium between the object and the lens.

At present, the smallest linear separation of two object points for which they may be resolved by an objective is fixed by the Rayleigh criterion:

$$d = 1.22(\lambda / 2NA)$$

where NA is the numerical aperture of the objective. Both the Abbe criteria and the Rayleigh criteria are very similar, being the numerical aperture related to the imaging medium by NA = nsinα. The maximal value of sinα is 1 (α = 90°), therefore the theoretical maximal numerical aperture of an objective in air (n = 1) is NA = 1. As a high NA is an essential requirement for high resolution, immersion optics has been developed. Biological specimens can be imaged at a very short distance from the objective through immersion media having a different refractive index, like water (n = 1.33), glycerine (n = 1.47), or oil (n = 1.52).

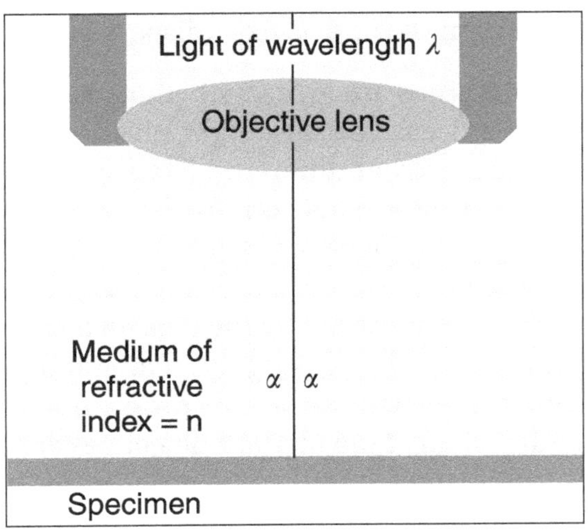

Factors appearing in the Rayleigh criterion: d = 1.22 (λ/2NA) = 1.22(λ/2 nsinα).

Optical Aberrations

Deviations from perfect image formation by optical systems are called aberrations. Unfortunately, thin lenses suffer from a variety of optical defects leading to distorted representations of the object, with the main final result of a rather severe loss of details from the observed specimen. Chromatic aberration is a very common optical defect related to the dependence of the refractive index of every material on the wavelength of light. The lens deviates the different color components of white light at different refraction angles, a phenomenon called dispersion, bringing them to focus at different points. This aberration gives rise to the formation of colored fringes surrounding the image. Manufacturers reduce or eliminate these problems by assembling objectives with a group of lenses having different dispersion properties (achromatic and apochromatic objectives).

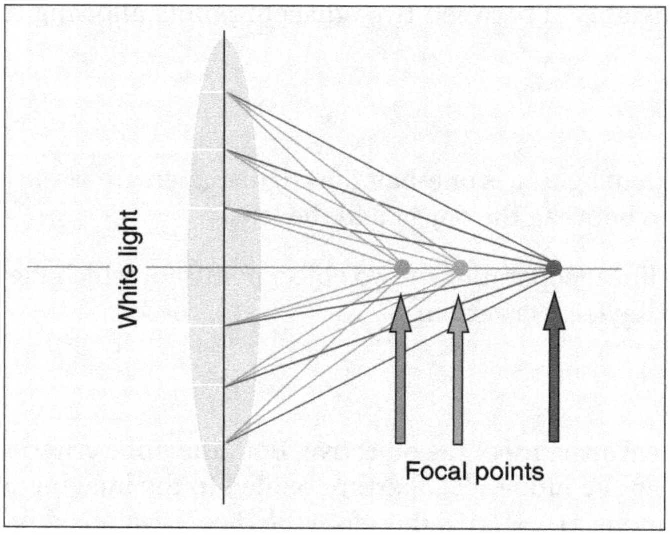

Chromatic aberration: The refraction index n depends inversely on the wavelength of incident light; therefore, thin lenses bend light differently as a function of wavelength. Lenses like the one shown in the figure will not image light of different color in exactly the same place because short wavelengths are refracted more than long wavelengths. The different color components of white light are focused at different points on the optical axis (dispersion of glass).

Spherical aberration and field curvature are artifacts both caused by the spherical curvature of lens surfaces. Spherical aberration occurs because light rays leaving the object at different distances from the optical axis are focalized at different distances along the axis when they are refracted by a lens with spherical surfaces. This serious defect, leading often to severe blurring of the images, may be reduced by using compound lenses in the objectives, made by sticking together convex and concave lenses with different thicknesses. Field curvature means that light rays originating from a flat surface are focalized by the lens on a concave spherical focal plane. It follows the impossibility of simultaneously focusing the center and the periphery of the flat specimen. Again, optical manufacturers partially solve this problem by inserting different groups of lenses in flat-field objectives. Almost all of the modern objectives are corrected for field curvature and are called plan.

Other major optical aberrations are coma (an error so that an object point has an asymmetrical image, a comet like spot), astigmatism (the failure of a lens to image a point as a single point, but as short segments that lie on two separated planes called sagittal and tangential focal plane), and geometrical distortion (straight lines are imaged as curved lines, so that the image of a square object may assume the aspect of a pincushion or a barrel). All these aberrations are largely corrected in modern objectives, but a careful alignment of the optical components of the microscope is an essential condition to minimize all these optical artifacts.

Contrast Enhancement

The formation of high-quality microscopic images strongly depends on the brightness and uniformity of specimen illumination. The most efficient system of illumination, both in transmitted light and in reflected light, was introduced by August Köhler in 1893. Unfortunately, unstained cells and tissues are "transparent" to light. The differences in optical density between a biological specimen and the background and among the different parts of the same specimen are insignificant;therefore, the very poor absorption of light does not allow the formation of an image that is contrasted enough. A good improvement can be obtained with dark field illumination. Blocking the light directly incident onto the specimen and allowing only the illumination by oblique light, bright images on a dark background can be obtained.

The possibility of taking advantage of the phase differences created among the light rays refracted from different parts of the specimen having a different refractive index and thickness. As a result of the slower speed of light in media with higher n, refracted light is retarded about one-quarter wavelength compared with the undeflected light. In 1938, Zernike built the first phase contrast microscope, designed to make visible differences in phase or optical path in transparent media, which is achieved by speeding up or slowing down the undeflected light of one-quarter wavelength, to obtain a phase difference with the refracted light of half a wavelength or zero wavelength. In these conditions, direct and refracted light interfere destructively or constructively at the eyepiece image plane (positive- or negative-phase contrast), converting differences of phase among rays coming from different points of the specimen into differences of amplitude, which can be visualized as differences in image contrast. Essential elements of a phase contrast microscope are an annular diaphragm in the front focal plane of the substage condenser and a phase plate, or "phase shifter", at the rear focal plane of the objective.

Another very efficient design for the enhancement of contrast in transparent specimens is the differential interference contrast (DIC). Light beams incident on the specimen are first polarized then splitted in an ordinary and an extraordinary ray by a birefringent Wollaston prism, modified by the French scientist Georges Nomarski (configuration called also Nomarski optics). Both rays pass through the specimen in very near regions, experiencing variations of their optical paths because of differences of refractive index and thickness. Modified rays are recombined by a second Wollaston prism and analyzed by a second polarizer. Light beams then interfere both constructively and destructively at the eyepiece image plane, showing details of the specimen having a different thickness as a separation between different colors. The overall effect is a pseudo-3-D image of the biological specimen.

Fluorescence Microscopy

Fluorescence is the emission of radiation in the visible range of the electromagnetic spectrum, by some materials stimulated by the absorption of another electromagnetic radiation having higher energy. Materials experiencing fluorescence are called fluorophores or fluorochromes and are characterized by an excitation and an emission spectrum.

An example of a fluorescent molecule: In the upper part of the figure, the structure of Fluorescein ($C_{20}H_{12}O_5$) is shown a molecule that exhibits intense yellow-green fluorescence in highly diluted solutions and is used, conjugated to antibodies, in biology and in medicine for diagnostic purposes. The normalized fluorescence excitation and emission spectra are shown in the lower part of the figure. The light absorption properties of a molecule are shaped by both the number of aromatic rings and the number of double bonds.

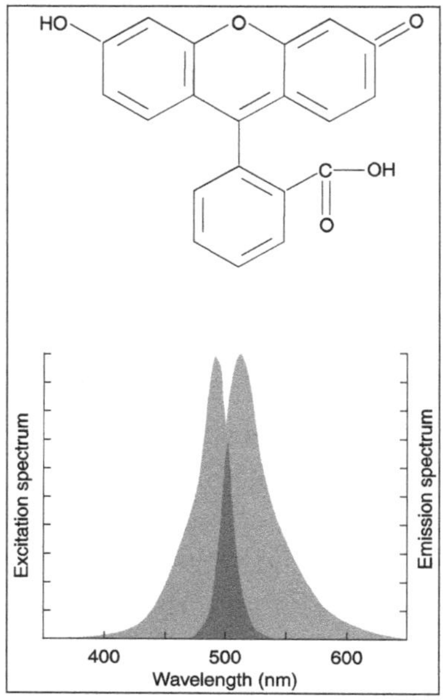

The fluorescence microscope is a modification of the compound optical microscope, arranged to admit light of a selected wavelength to the stained specimen and to image it through the light emitted by the fluorescent dyes. The light source of a fluorescent microscope is a gas-vapor arc lamp, using either a discrete (Mercury) or a continuous (Xenon) light-emission spectrum. Continuous- or pulsed-light-emitting lasers are now used in modern fluorescence microscopy tecniques, as confocal laser-scanning microscopy and multi-photon excitation microscopy. In the most common configuration (epifluorescence microscope), the excitation wavelength, selected by an excitation filter, is reflected by a suitable dichroic mirror (dichromatic beam splitter), reaching the specimen through the objective. A careful regulation of excitation intensity helps to avoid extinction of the fluorophore emission (photo-bleaching). The fluorescence light emitted by the fluorescent dyes is collected back by the objective, transmitted by the dichroic mirror, selected from undesired wavelengths by an emission filter (stop filter), and finally focalized at the eyepiece image plane.

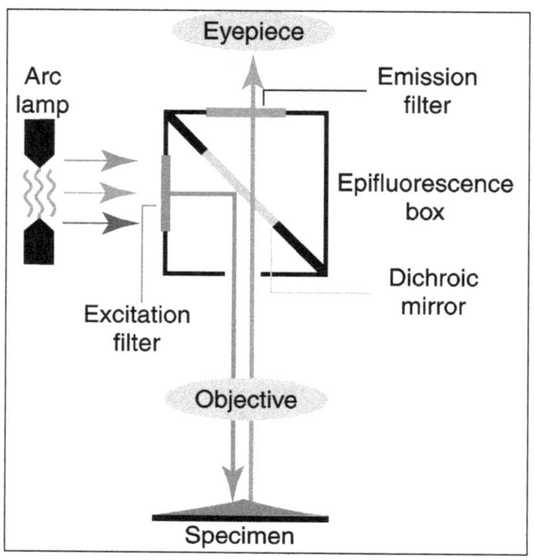

Schematic drawing of a conventional epifluorescence microscope.

The excitation filter selects the wanted wavelength from the different color components of the light generated by the gas-vapor arc lamp. Light, reflected by an appropriate dichroic mirror, hits the fluorescent specimen passing through the objective. Light emitted by the fluorescent dyes is collected back by the objective, transmitted by the dichroic mirror, selected from undesired wavelengths by the emission filter, and finally focalized at the eyepiece image plane. In most epifluorescence microscopes, different combinations of excitation filters, dichroic mirrors, and emission filters are pre-assembled in an epifluorescence box.

Advanced Techniques in Fluorescence Microscopy

In recent decades, some advanced techniques in fluorescence microscopy have been demonstrated to be especially suitable for biomedical applications. These techniques have been further strengthened and improved with the coming of confocal laser-scanning microscopy and multiphoton excitation microscopy.

Fluorescence Resonance Energy Transfer (FRET)

Fluorescence resonance energy transfer is a quantum-mechanical process described over 50 years ago by Theodor Förster. FRET is a distance-dependent dipole-dipole interaction between neighboring fluorophores, consisting in a non-radiative transfer of the excitation energy of the donor molecule to the ground state of an acceptor fluorophore. In the presence of FRET, excitation of the donor molecule will produce both quenching of the donor fluorescence and emission of light at the acceptor wavelength. FRET efficiency depends on the inverse sixth power of the donor-acceptor distance, making it suitable for investigating biological phenomena over distances on the order of the size of a typical protein. In a certain sense, FRET imaging may override the resolution limits of optical microscopy in applications like structure, conformational shifts, binding, colocalization, interactions between proteins, nucleic acids, and other biological macromolecules. Three main conditions are required for an optimal energy transfer:

- Donor emission spectrum and acceptor absorption spectrum should substantially overlap. The common spectral area, spectral overlap integral, must be at least 30% of the total.
- The distance between donor and acceptor molecules should typically range from 10 to 100 A°.
- The donor and acceptor transition dipoles must be in the proper orientation with respect to one another.

The upper part of the figure shows the excitation and emission spectra of donor and acceptor molecules. Donor emission spectrum and acceptor excitation spectrum are partially overlapped (dashed area). When the intermolecular distance is larger than 100 A°, the fluorescence properties of both molecules are not affected. As shown in the lower part of the figure, for intermolecular distances shorter than 100 A°, the excitation energy of donor is transferred to the acceptor. The donor decays to the ground state through a nonradiative energy loss and the excited acceptor through fluorescence emission. The common spectral area (spectral overlap integral), shown in black in the figure, must be at least 30% of the total. The donor and acceptor transition dipoles (white arrows) must be in the proper orientation with respect to one another.

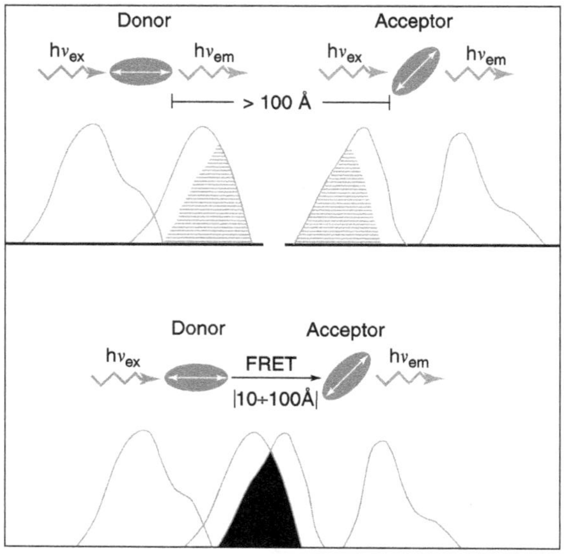

Fluorescence Lifetime Imaging (FLIM)

In the absence of molecular interactions, every fluorescent molecule is characterized by a unique duration of the excited state. Typical fluorescence decays are exponentials with a time-constant in the order of 10^{-9}s. Fluorescence lifetime is a peculiar property of a fluorophore that, unlike fluorescence emission intensity, is independent of photo bleaching, excitation light intensity, and optical path length in biological specimens. In FLIM, lifetime, rather than intensity, of the fluorescence emission is measured and spatially mapped in tissues and cells.

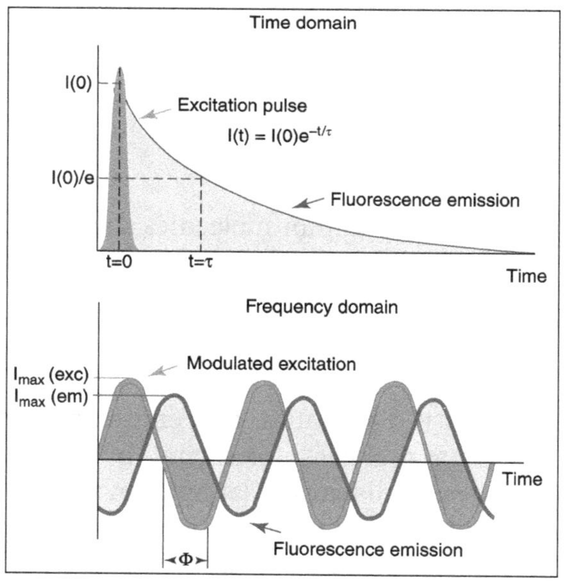

Fluorescence lifetime imaging: In the time-domain technique, fluorophores are excited at t = 0 by a very short light pulse (green area). The gradual decay of fluorescence intensity with time (red curve) can be fitted by a single-exponential function, according to I(t) = I(0) exp (-t/ τ), where τ is the fluorescence lifetime. In the frequency-domain technique, fluorophores are continuously excited with a sinusoidally modulated light (green curve). The emitted fluorescence (red curve) shows a similar waveform, but is modulated in amplitude and phase-shifted of an angle F from the excitation curve. Both amplitude modulation and phase-shift depend on the fluorescence emission lifetime τ.

Fluorescence lifetime may be measured according to the frequency-domain technique or the time-domain technique. In the time-domain technique, the fluorophore is excited by a very short pulse of light and the time-course of the emission intensity is monitored and recorded. Excitation pulses are usually generated by solid-state lasers or by mode-locked lasers with pulse duration of a few hundred picoseconds. The fluorescence image can be obtained by gated special detectors making use of image-intensification techniques. The pseudocolor representation of images at all pixels in the field of view is related to the fluorescence lifetime across the biological specimen. In the frequency-domain technique, the specimen is continuously excited with a sinusoidally modulated laser light (1–200 megahertz). The fluorescence lifetime can be measured from the phase shift between the excitation and the fluorescence and from the reduced emission amplitude modulated at the same frequency of excitation.

The lifetime of fluorescence emission is environmentally sensitive. Imaging using fluorescence lifetimes may provide information about functional properties of biological samples under investigation. Modifications induced by the environment in fluorescence lifetimes of fluorophores used as remote sensors allow one to measure ion concentrations (mainly oxygen, hydrogen, and calcium), hydrophobic properties, binding, and interactions among macromolecules in living cells.

Fluorescence Recovery after Photobleaching (FRAP)

Fluorescence recovery after photobleaching is a technique suitable for investigating the mobility of fluorescent molecules in living cells. Thanks to the introduction of green fluorescent proteins and other cell-labeling techniques, FRAP, born in the late 1970s to study viscosity of biological membranes and diffusional properties of molecules in lipid bilayers, is now a very popular modern imaging technique. In a FRAP experiment, a brief intense light pulse from a laser is used to irreversibly bleach fluorophores in a limited volume of a biological sample. If fluorescent molecules in the neighboring areas are almost partially mobile, fluorophores repopulate the bleached volume that regains a fluorescence signal. From the time-course and the extent of recovery of fluorescence intensity in the bleached volume, the translational diffusion coefficient and the mobile fraction of fluorophores may be calculated. In a related bleaching technique, named fluorescence loss in photobleaching (FLIP), a volume within the cell is subjected to repeated photobleaching, which will result in a reduction of the fluorescent signal from neighboring areas if the fluorescent molecules may diffuse through the bleached volume. The rate at which fluorescence is lost provides an indication of the mobility of fluorophores.

The algorithm, originally proposed by Axelrod et al. to solve the two-dimensional diffusion problems in biological membranes, was shown insufficient for three-dimensional molecular diffusion. Now, many different approaches allow one to identify and quantify three-dimensional diffusive phenomena in living cells, like motility of proteins and other macromolecules in cytoplasm, nucleus, and organelles or cytoskeletal dynamics.

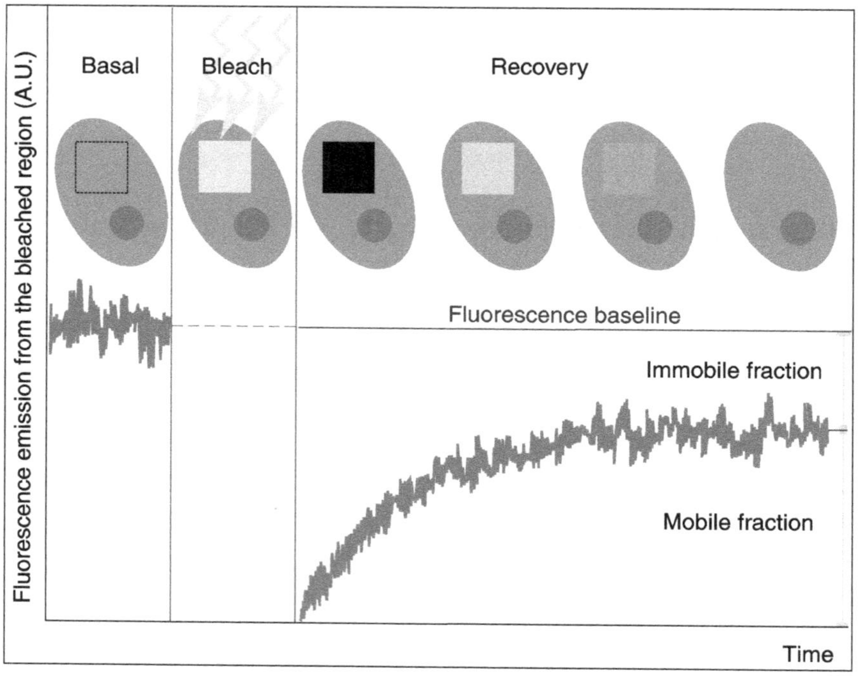

Fluorescence recovery after photobleaching: Graphical presentation of a FRAP experiment. A cell and the value of fluorescence intensity (fluorescence baseline) emitted before photobleaching is shown in the left side of the figure (basal). The fluorescence intensity of a region into the cell is significantly reduced by a very strong excitation (bleach). The fluorescence intensity in the photobleached area increases over time as unbleached molecules diffuse into this area (recovery). The fluorescence intensity almost never recovers to the baseline, the difference caused by a fraction of immobile fluorescent molecules. The lateral mobility is determined by the slope of the curve; the steeper the curve, the faster the recovery and, therefore, the more mobile the molecules.

Total Internal Reflection Fluorescence Microscopy (TIRFM)

According to Snell law, a light beam propagating from a medium of higher index of refraction n_1 to a medium of lower index of refraction n_2 does not undergo refraction if $\theta_1 = \theta_c = \sin^{-1}(n_2/n_1)$. θ_c is called critical angle, and for incident angles > θ_c the light beam is totally reflected. In a biological specimen, light crosses the interface between the glass coverslip and the aqueous medium being refracted and reflected as a function of the incident angle θ_g. The critical angle is $\theta_{c(g\rightarrow w)} = \sin^{-1}(n_w/n_g)$, where $n_w \cong 1.33$ and $n_g \cong 1.52$. If $\theta_g > \theta_{c(g\rightarrow w)}$ light is totally reflected from the glass/water interface, rather than passing through.

Although the light is totally reflected, a portion of the incident energy, named evanescent wave, penetrates through the interface and propagates in the aqueous medium. The intensity of this electromagnetic field decays very rapidly with the distance z from the interface, according to $I(z) = I(0)\, e^{-z/d}$, where d is a factor dependent on the beam wavelength, the refraction indices and the incident angle. The evanescent wave propagates in the aqueous medium only for less than 200 nanometers, however, having the same wavelength of the incident light, it is able to excite fluorophores in a very thin region of the biological specimen facing the interface.

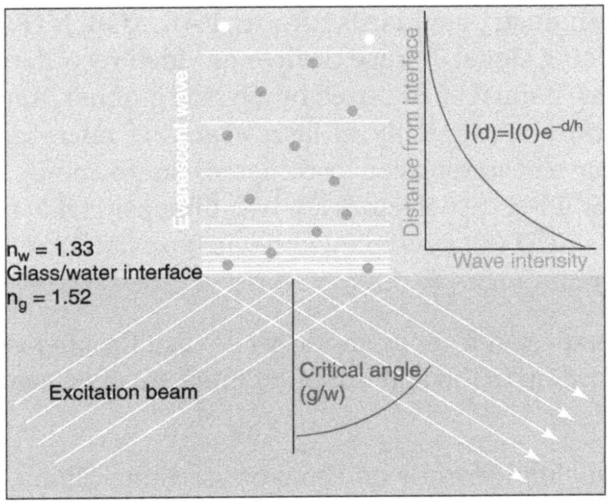

Schematic drawing of an excitation beam incident at a critical angle on the interface between a glass slide and an aqueous medium. Arrows show the direction of the beam. The light is totally reflected at the interface whereas the evanescent wave extends into the z direction. The intensity of evanescent wave decreases exponentially with the distance d according to I(d) = I(0) exp (-d/h). Fluorophores near the glass interface are excited by the evanescent wave and emit fluorescence (dark green spheres). The intensity of emitted fluorescence decreases with the intensity of the exciting evanescent wave (pale green spheres) and disappears in about 200 nanometers.

This phenomenon is the physical foundation of the total internal reflection fluorescence microscopy, a technique mainly used to observe fluorescence of molecules placed very near to the cell surface, and allowing to study, for instance, the cell-substrate contact, the adsorption of macromolecules to surfaces, and different cell-membrane processes. The main advantage of this technique is a result of the exponential decay of the evanescent wave, which prevents the excitation of fluorescence from outside the focal plane, which allows one to focus on an optical section about five times thinner than optical sections achieved with confocal microscopy, by far improving both the signal-to-noise ratio and the spatial resolution.

Many different optical configurations have been designed to realize TIRFM with a conventional epifluorescence microscope. They can be roughly divided in two main groups. The first group makes use of an added prism to direct a laser beam toward the glass/water interface with an angle wider than the critical angle. The second group uses immersion objectives with a very high numerical aperture to both direct illumination to the interface at supercritical angles and capture the emitted fluorescence.

Detection of Microscopic Images

For several decades, detection of microscopic images has been committed to ordinary photographic cameras using emulsion-based films as storage medium. Thanks to the rapid improvement of photographic technology, in the last twenty years electronic imaging has gradually replaced conventional photomicrography. Initially, cameras were video rate cameras. This type of camera makes use of light detectors based on vacuum-tube technology and generates an analog output signal that conforms to the industry standards RS-170, PAL, RGB, NTSC, etc. The electrical signal is usually recorded into analog signal storage devices as video recorders; however, to be sent to a computer for image analysis, it must be digitized by a frame grabber. Actually, vacuum-tube-based detectors (photomultipliers) are used only in laser-scanning microscopy techniques. In recent years, video rate cameras have been replaced by cameras using a solid-state detector whose output is already a digital signal. Solid-state detectors are two-dimensional arrays of photodiodes mainly manufactured according to CCD (charge-coupled device) or CMOS (complimentary metal-oxide semiconductor) technology.

High-level performance is expected from scientific grade digital cameras, but the device properties heavily affect its price. According to the application of interest, a few main features must be considered:

- Resolution, which mainly depends on the cross-sectional illuminated area of every photodiode (pixel). Smaller pixels and the resulting higher pixel density in the array grant better resolution.
- Sensitivity is affected by the exposure time, the pixel area, and the quantum efficiency of the CCD. Many different strategies have been applied to maximize sensitivity (back-illuminated cameras, intensified cameras, etc).
- Signal-to-noise ratio, the main source of noise in a digital camera is the readout noise, dependent on the CCD manufacturing, the electronics of the device, and the readout speed. The effect of thermal noise can be reduced by cooling the camera.
- Dynamic range is the ratio between the maximum light intensity observable without saturation of the CCD and the noise level, and is affected by pixel size and device architecture. The dynamic range influences the number of gray or color values that can be measured and digitized; typically, an 8-bit (256 levels) or 12-bit (4096) digital representation is adopted.

Optical Telescope

An optical telescope is a telescope which is used to gather, and focus light, for directly viewing a magnified image, making a photograph, etc. The term is used especially for a monocular with static mounting for observing the sky. Handheld binoculars are common for other purposes. Light is made up of photons, and professional telescopes concentrate the light onto electronic detectors which collect the photons. There are three primary types of optical telescope: Refractors (Dioptrics) which use lenses, Reflectors (Catoptrics) which use mirrors, and Combined Lens-Mirror Systems (Catadioptrics) which use lenses and mirrors in combination. The main tasks of a telescope are the following:

- Gathering as much light radiation as possible.
- Increasing an angular separation between objects.
- Creating a focused image of an object.

We have now achieved high technical level, which enables us to create colossal telescopes, reaching distant regions of the Universe and making great discoveries.

Telescope Components

The main parts of which any telescope consists with are the following:

- Primary lens (for refracting telescopes), which is the main component of a device. Bigger the lens, more light a telescope can gather and fainter objects can be viewed.
- Primary mirror (for reflecting telescopes), which carries the same role as the primary lens in a refracting telescopes.
- Eyepiece, which magnifies the image.
- Mounting, which supports the tube, enabling it to be rotated.

Refracting Telescopes

The history of refractor's invention was discussed in the introduction part. Refractor is the simplest type of telescopes, which combines two lenses at the ends of a tube. As mentioned above, the main component of this type of telescope is the primary lens – the objective. A concave lens is used as an objective of a refractor. It defines how faint objects can be viewed. The eyepiece, another concave lens, is placed at the other end of the tube. It defines the magnification of an object observed. The appearance and a schematic view of refracting telescope are shown in Figure.

The refracting telescope.

Reflecting Telescope

This is a later type of telescopes, firstly proposed by Giovanni Francesco Sagredo who suggested that a curved mirror could be used instead of a lens. The first reflecting prototype was created by Isaac Newton in 1669. These telescopes employ parabolic mirror to gather light from distant space sources and several auxiliary mirrors which direct path to the eyepiece. The appearance and a schematic view of reflecting telescope (Newtonian design) are shown in figure.

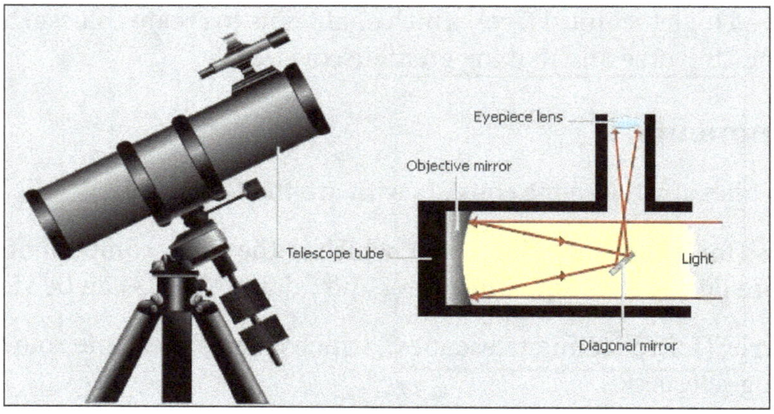

The reflecting telescope.

Principle of Working and Main Parameters

The light from a source reaches the primary lens (or primary mirror), converging at a focal point of a system. Depending on the design, the light either enters the eyepiece (refractors) or is reflected outside a main tube towards the ocular lens (reflectors). The angular resolution is defined by the diameter of the main lens or mirror, which is referred to as an aperture D, measured in millimeters. For the visible light this limit is described by the following formula:

$$\sin\alpha = \frac{138}{D}$$

where α is the limit resolution measured in arc seconds.

Each lens or mirror has their special feature known as focal length. If the focal length of the objective is F and the focal length of the eyepiece is f, then the magnification of the image produced by the optics is described by the following simple formula:

$$M = \frac{F}{f}$$

The bigger the aperture, more light the optics can gather. More light enters the telescope, the fainter objects can be viewed. Limiting magnitude of a telescope can be found according the following formula:

$$M_L \approx 2.7 + 5\log d$$

where d is the aperture diameter in millimeters.

Another important parameter of a telescope is the focal ratio. It is defined as the focal length of the objective divided by its diameter. The smaller the ratio, the "faster" the optical system is and the better and brighter image can be produced.

Mountings

There are two main classes of mount: altitude-azimuth and equatorial. Telescopes with the first mount are aligned with the local zenith. Experiencing symmetrical gravity forces, the telescopes are normally less massive. The disadvantage is that the device should be moved in two directions simultaneously in order to track the object. Equatorial mount is aligned with the polar axis. It requires a heavy counterweight to balance the telescope. The advantage of this mount is that the tube has to be rotated along one polar axis only for following the sky. This configuration is much more convenient for tracking objects.

Optical Defects

The quality of the views produced by optical telescopes is limited by optical defects - aberrations. Refractors suffer from chromatic aberration more than from other deviations. This is the result of the difference of the speed of light of various wavelengths in medium. Red light is refracted more than blue one.

As the consequence of this effect, a star is rounded by colorful concentric rings of light. A corrective lens is placed on top of the primary objective to reduce chromatic aberration. Other types of aberrations are monochromatic. They are spherical, coma, astigmatism, distortion and field curvature. Spherical aberration results from the special spherical surface feature – it focuses on a line, not a point. Therefore, rays from the center of a mirror focus farther from it than rays from the edges. This is fixed by employing parabolic mirrors. The rest of aberrations are off-axis, they depend on the field angle.

Optical Periscope

Periscope is an optical instrument used in land and sea warfare, submarine navigation, and elsewhere to enable an observer to see his surroundings while remaining under cover, behind armour, or submerged. A periscope includes two mirrors or reflecting prisms to change the direction of the light coming from the scene observed: the first deflects it down through a vertical tube, the second diverts it horizontally so that the scene can be viewed conveniently. Frequently there is a telescopic optical system that provides magnification, gives as wide an arc of vision as possible, and includes a crossline or reticle pattern to establish the line of sight to the object under observation. There may also be devices for estimating the range and course of the target in military applications and for photographing through the periscope.

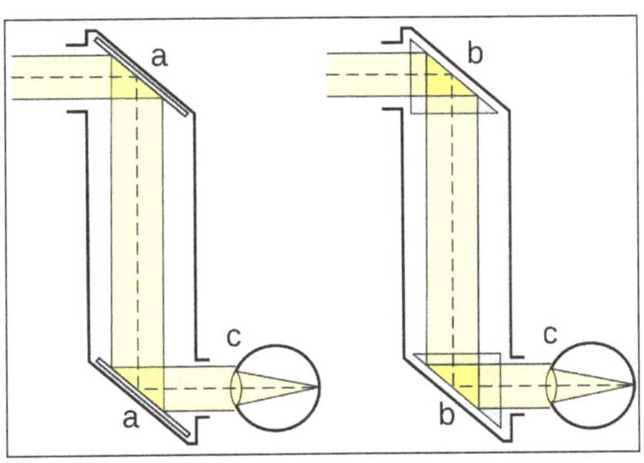

Periscope, eyepiece box and observer's station; handles control rotation about the axis, twist grips provide control of the line-of-sight elevation.

The simplest type of periscope consists of a tube at the ends of which are two mirrors, parallel to each other but at 45° to the axis of the tube. This device produces no magnification and does not give a crossline image. The arc of vision is limited by the simple geometry of the tube: the longer or narrower the tube, the smaller the field of view. Periscopes of this type were widely used in World War II in tanks and other armoured vehicles as observation devices for the driver, gunner, and commander. When fitted with a small, auxiliary gunsight telescope, the tank periscope can also be used in pointing and firing the guns. By employing tubes of rectangular cross section, wide, horizontal fields of view can be obtained.

Optical Computing

Optics has been used in computing for a number of years but the main emphasis has been and continues to be to link portions of computers, for communications, or more intrinsically in devices that have some optical application or component (optical pattern recognition, etc.). Optical digital computers are still some years away, however a number of devices that can ultimately lead to real optical computers have already been manufactured, including optical logic gates, optical switches, optical interconnections, and optical memory. The most likely near-term optical computer will really be a hybrid composed of traditional architectural design along with some portions that can perform some functional operations in optical mode.

With today's growing dependence on computing technology, the need for high performance computers (HPC) has significantly increased. Many performance improvements in conventional computers are achieved by miniaturizing electronic components to very small micron-size scale so that electrons need to travel only short distances within a very short time. This approach relies on the steadily shrinking trace size on microchips (i.e., the size of elements that can be 'drawn' onto each chip). This has resulted in the development of Very Large Scale Integration (VLSI) technology with smaller device dimensions and greater complexity. The smallest dimensions of VLSI nowadays are about 0.08 mm. Despite the incredible progress in the development and refinement of the basic technologies over the past decade, there is growing concern that these technologies may not be capable of solving the computing problems of even the current millennium. Applications of HPC and visualization technologies lead to breakthroughs in engineering and manufacturing in a wide range of industries. With the help of virtual product design and development, costs can be reduced; hence looking for improved computing capabilities is desirable. Optical computing includes the optical calculation of transforms and optical pattern matching. Emerging technologies also make the optical storage of data a reality.

The speed of computers was achieved by miniaturizing electronic components to a very small micron-size scale, but they are limited not only by the speed of electrons in matter (Einstein's principle that signals cannot propagate faster than the speed of light) but also by the increasing density of interconnections necessary to link the electronic gates on microchips. The optical computer comes as a solution of miniaturization problem. In an optical computer, electrons are replaced by photons, the subatomic bits of electromagnetic radiation that make up light. Optics, which is the science of light, is already used in computing, most often in the fiber-optic glass cables that currently transmit: data on communication networks much faster than via traditional copper wires.

Thus, optical signals might be the ticket for the fastest supercomputers ever. Compared to light, electronic signals in chips travel at snail speed. Moreover, there is no such thing as a short circuit with light, so beams could cross with no problem after being redirected by pinpoint-size mirrors in a switchboard. In a pursuit to probe into cutting-edge research areas, optical technology (optoelectronic, photonic devices) is one of the most promising, and may eventually lead to new computing applications as a consequence of faster processor speeds, as well as better connectivity and higher bandwidth.

The pressing need for optical technology stems from the fact that today's' computers are limited by the time response of electronic circuits. A solid transmission medium limits both the speed and volume of signals, as well as building up heat that damages components. For example, a one-foot length of wire produces approximately one nanosecond (billionth of a second) of time delay. Extreme miniaturization of tiny electronic components also leads to 'cross-talk' - signal errors that affect the system's reliability. These and other obstacles have led scientists to seek answers in light itself. Light does not have the time response limitations of electronics, does not need insulators, and can even send dozens or hundreds of photon signal streams simultaneously using different color frequencies. Those are immune to electromagnetic interference, and free from electrical short circuits. They have low-loss transmission and provide large bandwidth; i.e. multiplexing capability, capable of communicating several channels in parallel without interference. They are capable of propagating signals within the same or adjacent fibers with essentially no interference or cross talk. They are compact, lightweight, and inexpensive to manufacture, as well as more facile with stored information than magnetic materials. By replacing electrons and wires with photons, fiber optics, crystals, thin films and mirrors, researchers are hoping to build a new generation of computers that work 100 million times faster than today's machines.

Optical computing was a hot research area in the 1980s. But the work tapered off because of materials limitations that seemed to prevent optochips from getting small enough and cheap enough to be more than laboratory curiosities. Now, optical computers are back with advances in self-assembled conducting organic polymers that promise super-tiny all-optical chips. Advances in optical storage device have generated the promise of efficient, compact and large-scale storage devices. Another advantage of optical methods over electronic ones for computing is that parallel data processing can frequently be done much more easily and less expensively in optics than in electronics.

Parallelism, the capability to execute more than one operation simultaneously, is now common in electronic computer architectures. But, most electronic computers still execute instructions sequentially; parallelism with electronics remains sparsely used. Its first widespread appearance was in Cray supercomputers in the early 1980's when two processors were used in conjunction with one shared memory. Today, large supercomputers may utilize thousands of processors but communication overhead frequently results in reduced overall efficiency. On the other hand for some applications in input-output (I/O), such as image processing, by using a simple optical design an array of pixels can be transferred simultaneously in parallel from one point to another.

Optical technology promises massive upgrades in the efficiency and speed of computers, as well as significant shrinkage in their size and cost. An optical desktop computer could be capable of processing data up to 100,000 times faster than current models because multiple operations can be performed simultaneously. Other advantages of optics include low manufacturing costs, immunity to electromagnetic interference, a tolerance for low-loss transmissions, freedom from short electrical circuits and the capability to supply large bandwidth and propagate signals within the same or adjacent fibers without interference. One oversimplified example may help to appreciate the difference between optical and electronic parallelism. Consider an imaging system with 1000 x 1000 independent points per mm^2 in the object plane which are connected optically by a lens to a corresponding number of points per mm^2 in the image plane; the lens effectively performs an FFT of the image plane in real time. For this to be accomplished electrically, a million operations are required.

Parallelism, when associated with fast switching speeds, would result in staggering computational speeds. Assume, for example, there are only 100 million gates on a chip, much less than what was mentioned earlier (optical integration is still in its infancy compared to electronics). Further, conservatively assume that each gate operates with a switching time of only 1 nanosecond (organic optical switches can switch at sub-picosecond rates compared to maximum picosecond switching times for electronic switching). Such a system could perform more than 10^{17} bit operations per second. Compare this to the gigabits (10^9) or terabits (10^{12}) per second rates which electronics are either currently limited to, or hoping to achieve. In other words, a computation that might require one hundred thousand hours (more than 11 years) of a conventional computer time could require less than one hour by an optical one.

But building an optical computer will not be easy. A major challenge is finding materials that can be mass produced yet consume little power; for this reason, optical computers may not hit the consumer market for 10 to 15 years. Another of the typical problems optical computers have faced is that the digital optical devices have practical limits of eight to eleven bits of accuracy in basic operations due to, e.g., intensity fluctuations. Thus, for example, digital partitioning algorithms, that can break matrix-vector products into lower-accuracy sub-products, working in tandem with error-correction codes, can substantially improve the accuracy of optical computing operations.

Nevertheless, many problems in developing appropriate materials and devices must be overcome before digital optical computers will be in widespread commercial use. In the near term, at least, optical computers will most likely be hybrid optical/electronic systems that use electronic circuits to preprocess input data for computation and to post-process output data for error correction before outputting the results. The promise of all optical computing remains highly attractive, however, and the goal of developing optical computers continues to be a worthy one. Nevertheless, many scientists feel that an all optical computer will not be the computer of the future; instead optoelectronic computers will rule where the advantages of both electronics and optics will be used. Optical computing can also be linked intrinsically to quantum computing. Each photon is a quantum of a wave function describing the whole function. It is now possible to control atoms by trapping single photons in small, superconducting cavities. So photon quantum computing could become a future possibility.

Key Optical Components for Computing

The major breakthroughs on optical computing have been centered on the development of micro-optic devices for data input. Conventional lasers are known as 'edge emitters' because their laser light comes out from the edges. Also, their laser cavities run horizontally along their length. A vertical cavity surface emitting laser (VCSEL - pronounced 'vixel'), however, gives out laser light froQm its surface and has a laser cavity that is vertical; hence the name.

VCSEL is a semiconductor vertical cavity surface emitting micro laser diode that emits light in a cylindrical beam vertically from the surface of a fabricated wafer, and offers significant advantages when compared to the edge-emitting lasers currently used in the majority of fiber optic communications devices. They emit at 850 nm and have rather low thresholds (typically a few mA). They are very fast and can give mW of coupled Power into a 50 micron core fiber and are extremely radiation hard. VCSELS can be tested at the wafer level (as opposed to edge emitting lasers which have to be cut and cleaved before they can be tested) and hence are relatively cheap. In fact, VCSELs can be fabricated efficiently on a 3-inch diameter wafer. A schematic of VCSEL is shown in figure.

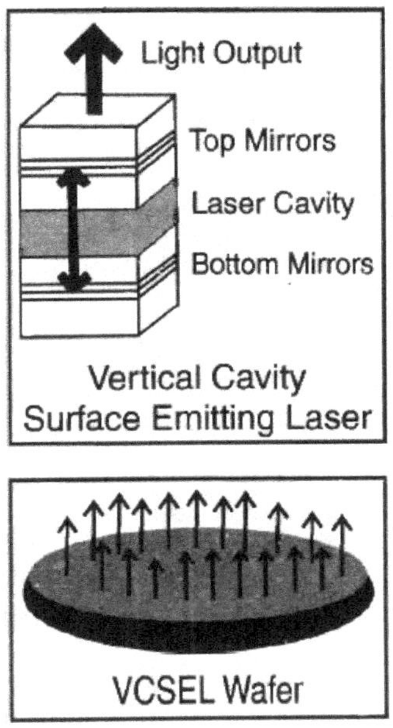

A schematic of the VCSSEL and its possible miniaturization into wafer.

The principals involved in the operation of a VCSEL are very similar to those of regular lasers. As shown in Figure, there are two special semiconductor materials sandwiching an active layer where all the action takes place. But rather than reflective ends, in a VCSEL there are several layers of partially reflective mirrors above and below the active layer. Layers of semiconductor with differing compositions create these mirrors, and each mirror reflects a narrow range of wavelengths back into the cavity in order to cause light emission at just one wavelength.

Spatial light modulators (SLMs) play an important role in several technical areas where the control of light on a pixel-by-pixel basis is a key element, such as optical processing, for inputting information on light beams, and displays. For display purposes the desire is to have as many pixels as possible in as small and cheap a device as possible. For such purposes designing silicon chips for use as spatial light modulators has been effective. The basic idea is to have a set of memory cells laid out on a regular grid. These cells are electrically connected to metal mirrors, such that the voltage on the mirror depends on the value stored in the memory cell. A layer of optically active liquid crystal is sandwiched between this array of mirrors and a piece of glass with a conductive coating. The voltage between individual mirrors and the front electrode affects the optical activity of the liquid crystal in that neighborhood. Hence by being able to individually program the memory locations one can set up a pattern of optical activity in the liquid crystal layer. Figure (a) shows a reflective 256 x 256 pixel device based on SRAM technology. Several technologies have contributed to the development of SLMs. These include micro-electro-mechanical devices, such as, acoustic optic modulators (AOMs), and pixelated electro optical devices, such as liquid-crystal modulators (LCMs). Figure (b) shows a simple AOM operation in deflecting light beam direction. Encompassed within these categories are amplitude only, phase-only, or amplitude-phase modulators.

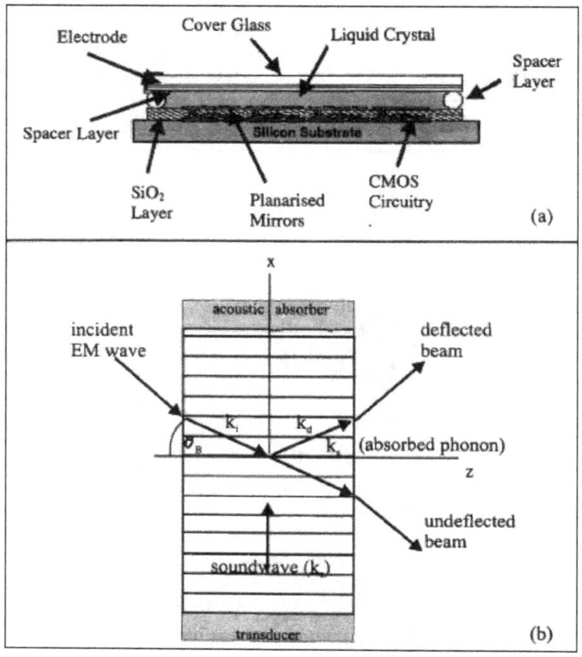

Figure above shows: (a) A reflective 256x256 pixel device based on SRAM (Static Random Access Memory) technology with embedded CMOS (Complimentary Metal-Oxide Semiconductor). (b) Schematic diagram of an Acousto-Optic Modulator (AOM) deflection principle.

Broadly speaking, an optical computer is a computer in which light is used somewhere. This can means fiber optical connections between electronic components, free space connections, or one in which light functions as a mechanism for storage of data, logic or arithmetic. Instead of electrons in silicon integrated circuits, the digital optical computers will be based on photons. Smart pixels, the union of optics and electronics, both expands the capabilities of electronic systems and enables optical systems with high levels of electronic signal processing. Thus, smart pixel systems add value to electronics through optical input/output and interconnection, and value is added to optical systems through electronic enhancements which include gain, feedback control, and image processing and compression.

Smart pixel technology is a relatively new approach to integrating electronic circuitry and optoelectronic devices in a common framework. The purpose is to leverage the advantages of each individual technology and provide improved performance for specific applications. Here, the electronic circuitry provides complex functionality and programmability while the optoelectronic devices provide high-speed switching and compatibility with existing optical media. Arrays of these smart pixels leverage the parallelism of optics for interconnections as well as computation. A smart pixel device, a light emitting diode (LED) under the control of a field-effect transistor (FET), can now be made entirely out of organic materials on the same substrate for the first time.

In general, the benefit of organic over conventional semiconductor electronics is that they should (when mass-production techniques take over) lead to cheaper, lighter, circuitry that can be printed rather than etched. Scientists at Bell Labs have made 300-micron-wide pixels using polymer FETs and LEDs made from a sandwich of organic materials, one of which allows electrons to flow, another which acts as highway for holes (the absence of electrons); light is produced when electrons and holes meet. The pixels are quite potent, with a brightness of about 2300 candela/m^2, compared to a figure of 100 for present flat-panel displays.

Uses of Optics in Computing

Currently, optics is used mostly to link portions of computers, or more intrinsically in devices that have some optical application or component. For example, much progress has been achieved, and optical signal processors have been successfully used, for applications such as synthetic aperture radars, optical pattern recognition, optical image processing, fingerprint enhancement, and optical spectrum analyzers. The early work in optical signal processing and computing was basically analog in nature. In the past two decades, however, a great deal of effort has been expended in the development of digital optical processors.

Much work remains before digital optical computers will be widely available commercially, but the pace of research and development has increased through the 1990s. During the last decade, there has been continuing emphasis on the following aspects of optical computing:

- Optical tunnel devices are under continuous development varying from small caliber endoscopes to character recognition systems with multiple type capability.
- Development of optical processors for asynchronous transfer mode.
- Development architectures for optical neural networks.
- Development of high accuracy analog optical processors, capable of processing large amounts of data in parallel.

Since photons are uncharged and do not interact with one another as readily as electrons, light beams may pass through one another in full-duplex operation, for example without distorting the information carried. In the case of electronics, loops usually generate noise voltage spikes whenever the electromagnetic fields through the loop changes. Further, high frequency or fast switching pulses will cause interference in neighboring wires. On the other hand, signals in adjacent optical fibers or in optical integrated channels do not affect one another nor do they pick up noise due to loops. Finally, optical materials possess superior storage density and accessibility over magnetic materials.

The field of optical computing is progressing rapidly and shows many dramatic opportunities for overcoming the limitations described earlier for current electronic computers. The process is already underway whereby optical devices have been incorporated into many computing systems. Laser diodes as sources of coherent light have dropped rapidly in price due to mass production. Also, optical CD-ROM discs are now very common in home and office computers.

Current trends in optical computing emphasize communications, for example the use of free-space optical interconnects as a potential solution to alleviate bottlenecks experienced in electronic architectures, including loss of communication efficiency in multiprocessors and difficulty of scaling down the IC technology to sub-micron levels. Light beams can travel very close to each other, and even intersect, without observable or measurable generation of unwanted signals. Therefore, dense arrays of interconnects can be built using optical systems. In addition, risk of noise is further reduced, as light is immune to electromagnetic interferences.

Finally, as light travels fast and it has extremely large spatial bandwidth and physical channel density, it appears to be an excellent media for information transport and hence can be harnessed for data processing. This high bandwidth capability offers a great deal of architectural advantage and flexibility. Based on the technology now available, future systems could have 1024 smart pixels per chip with each channel clocked at 200MHz (a chip I/O of 200Gbits per second), giving aggregate data capacity in the parallel optical highway of more than 200Tbits per second; this could be further increased to 1000Tbits. Free-space optical techniques are also used in scalable crossbar systems, which allow arbitrary interconnections between a set of inputs and a set of outputs. Optical sorting and optical crossbar inter-connects are used in asynchronous transfer modes or packet routing and in shared memory multiprocessor systems.

In optical computing two types of memory are discussed. One consists of arrays of one-bit-store elements and the other is mass storage, which is implemented by optical disks or by holographic storage systems. This type of memory promises very high capacity and storage density. The primary benefits offered by holographic optical data storage over current storage technologies include significantly higher storage capacities and faster read-out rates. This study is expected to lead to compact, high-capacity, rapid- and random-access, radiation-resistant, low-power, and low-cost data storage devices necessary for future intelligent spacecraft, as well as to massive-capacity and fast-access terrestrial data archives.

As multimedia applications and services become more and more prevalent, entertainment and data storage companies are looking at ways to increase the amount of stored data and reduce the time it takes to get that data out of storage. The SLMs and the linear array beam steered are used in optical data storage applications. These devices are used to write data into the optical storage medium at high speed. The analog nature of these devices means that data can be stored at much higher density than data written by conventional devices. Researchers around the world are evaluating a number of inventive ways to store optical data while improving the performance and capacity of existing optical disk technology. While these approaches vary in materials and methods, they do share a common objective: expanded capacity through stacking layers of optical material. For audio recordings, an lSo-MB minidisk with a 2.S-in. diameter has been developed that uses special compression to shrink a standard CD's 640-MB storage capacity onto the smaller polymer substrate. It is rewritable and uses magnetic field modulation on optical material. The mini disk uses one of two methods to write information onto an optical disk.

With the minidisk, a magnetic field placed behind the optical disk is modulated while the intensity of the writing laser head is held constant. By switching the polarity of the magnetic field while the laser creates a state of flux in the optical material, digital data can be recorded on a single layer. As with all optical storage media, a read laser retrieves the data. Along with mini disk developments, standard magneto-optical CD technology has expanded the capacity of the 3.S-in. diameter disk from 640 MB to commercially available 1 GB storage media. These conventional storage media modulate the laser instead of the magnetic field during the writing process. Fourth-generation 8 × 5.25 in. diameter disks that use the same technology have reached capacities of 4 GB per disk.

These disks are used mainly in 'jukebox' devices. These machines contain multiple disks for storage and backup of large amounts of data that need to be accessed quickly.

Beyond these existing systems are several laboratory systems that use multiple layers of optical material on a single disk. The one with the largest capacity, magnetic super-resolution (MSR), uses two layers of optical material. The data is written onto the bottom layer through a writing laser and magnetic field modulation (MFM). When reading the disk in MSR mode, the data is copied from the lower layer to the upper layer with greater spacing between bits. In this way, data can be stored much closer together (at distances smaller than the read beam wavelength) on the bottom layer without losing data due to averaging across bits. This method is close to commercial production, offering capacities of up to 20 GB on a 5.25 in. disk without the need for altering conventional read-laser technology.

Advanced storage magnetic optics (ASMO) builds on MSR, but with one exception. Standard optical disks, including those used in MSR, have grooves and lands just like a phonograph record. However, standard systems only record data in the grooves, not on the lands, wasting a certain amount of the optical material's capacity. ASMO records data on both lands and grooves and, by choosing groove depths approximately 1/6 the wavelength of the reading laser light, the system can eliminate the cross track crosstalk that would normally be the result of recording on both grooves and lands. Even conventional CD recordings pick up data from neighboring tracks, but this information is filtered out, reducing the signal-to-noise ratio. By closely controlling the groove depth, ASMO eliminates this problem while maximizing the signal-to-noise ratio. MSR and ASMO technologies are expected to produce removable optical disk drives with capacities between 6 and 20 GB on a 12-cm optical disk, which is the same size as a standard CD that holds 640 MB. Magnetic amplifying magneto-optical systems (MAMMOS) use a standard polymer disk with two or three magnetic layers. In general terms, MAMMOS is similar to MSR, except that when the data is copied from the bottom to the upper layer, it is expanded in size, amplifying the signal.

Technology developed by Call/Recall Inc. (San Diego, CA) could help bridge the gap between optical disk drives and holographic memories. Called 2-photon optical storage technology, the Call-1Recall systems under development use a single beam to write the data in either optical disk with up to 120 layers, or into 100-layer cubes of active-molecule-doped MMA polymer. In operation, a mask representing data is illuminated by a mode-locked Nd: YAG laser emitting at 1064 nm with pulse durations of 35 ps. The focal point of the beam intersects a second beam formed by the second harmonic of the same beam at 532 nm. The second beam fixes the data spatially and temporally.

A third beam from a HeNe laser emitting at 543 nm reads the data by causing the material to fluoresce. The fluorescence is read by a charge coupled device (CCD) chip and converted through proprietary algorithms back into data. Newer versions of the system use a Ti: Sapphire laser with 200-fs pulses. Call/Recall's Fredrick McCormick said the newer and older approaches offer different strengths. The YAG system can deliver higher-power pulses capable of storing megabits of data with a single pulse, but at much lower repetition rates than the Ti:Sapphire laser with its lower-power pulses. Thus, it is a trade-off. Call1Recall has demonstrated the system using portable apparatus comprised of a simple stepper-motor-driven stage and 200-microwatt HeNe laser in conjunction with a low-cost video camera. The company estimates that an optimized system could produce static bit error rates (BER) of less than 9×10^{-13}.

McCormick believes that a final prototype operating at standard CD rotation rates would offer BERs that match or slightly exceed conventional optical disk technology. Researchers such as Demetri Psaltis and associates at the California Institute of Technology are also using active-molecule-doped polymers to store optical data holographically. Their system uses a thin polymer layer of PMMA doped with phenanthrenequinone (PQ). When illuminated with two coherent beams, the subsequent interference pattern causes the PQ molecules to bond to the PMMA host matrix to a greater extent in brighter areas and to a lesser extent in areas where the intensity drops due to destructive interference. As a result, a pair of partially offsetting index gratings is formed in the PMMA matrix.

After writing the hologram into the polymer material, the substrate is baked, which causes the remaining unbounded PQ molecules to diffuse throughout the polymer, removing the offsetting grating and leaving the hologram. A uniform illumination is the final step, bonding the diffuse PQ throughout the matrix and fixing the hologram in the polymer material. Storagetek's Archie Smith estimates that devices based on this method could hold between 100 and 200 GB of data on a S.2S-in diameter polymer disk.

More conventional approaches to holographic storage use iron-doped lithium niobate crystals to store pages of data. Unlike standard magneto-optical storage devices, however, the systems developed by Pericles Mitkas at Colorado State University use the associative search capabilities of holographic memories. Associative or content-based data access enables the search of the entire memory space in parallel for the presence of a keyword or search argument. Conventional systems use memory addresses to track data and retrieve the data at that location when requested. Several applications can benefit from this mode of operation including management of large multimedia databases, video indexing, image recognition, and data mining. Different types of data such as formatted and unformatted text, gray scale and binary images, video frames, alphanumeric data tables, and time signals can be interleaved in the same medium and we can search the memory with either data type.

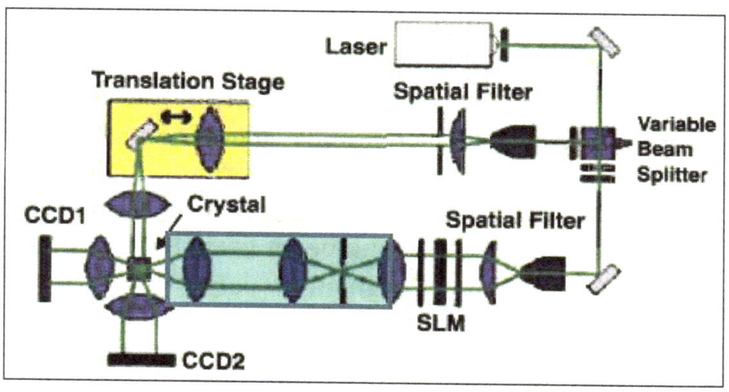

Holographic memory cubes use a spatial light modulator to simultaneously search the entire memory for a searchable object - be it text, image, or something else. This associative memory search process promises significant benefits for database searching and other applications.

The system uses a data and a reference beam to create a hologram on one plane inside the lithium niobate. By changing the angle of the reference beam, more data can be written into the cube just like pages in a book. The current systems have stored up to 1000 pages per spatial location in either VGA or VGA resolutions. To search the data, a binary or analog pattern that represents the search argument is loaded into a spatial light modulator and modulates a laser beam. The light diffracted by the holographic cube on a CCD generates a signal that indicates the pages that match the sought data. Recent results have shown the system can find the correct data 7S percent of the time when using patterns as small as 1 to 5 percent of the total page. That level goes up to 95 to 100 percent by increasing the amount of data included in the search argument.

References

- What-is-optical-fiber, physics: byjus.com, Retrieved 09, February 2020
- Fiber-Optics-Understanding-the-Basics-25151: photonics.com, Retrieved 12, May 2020
- Fiber-optics-and-its-importance: shireeninc.com, Retrieved 24, August 2020
- Optical-Microscopy-277704161: researchgate.net, Retrieved 06, April 2020
- Periscope, technology: britannica.com, Retrieved 10, January 2020

5

Optical Communication Systems

The means of communication which uses light for transmitting information is known as optical communication. The systems which are used for transmitting information optically through fibers by converting the electrical signal into light pulses are known as optical communication systems. The topics elaborated in this chapter will help in gaining a better perspective about optical communication systems, optical fibers and their applications.

Optical Communication

A naval signal lamp, a form of optical communication that uses shutters and is typically employed with Morse code (2002).

Optical communication, also known as optical telecommunication, is communication at a distance using light to carry information. It can be performed visually or by using electronic devices. The earliest basic forms of optical communication date back several millennia, while the earliest electrical device created to do so was the photophone, invented in 1880.

An optical communication system uses a transmitter, which encodes a message into an optical signal, a channel, which carries the signal to its destination, and a receiver, which reproduces the message from the received optical signal. When electronic equipment is not employed the 'receiver' is a person visually observing and interpreting a signal, which may be either simple (such as the presence of a beacon fire) or complex (such as lights using color codes or flashed in a Morse code sequence).

Free-space optical communication has been deployed in space, while terrestrial forms are naturally limited by geography, weather and the availability of light. This article provides a basic introduction to different forms of optical communication.

Forms

Visual techniques such as smoke signals, beacon fires, hydraulic telegraphs, ship flags and semaphore lines were the earliest forms of optical communication. Hydraulic telegraph semaphores date back to the 4th century BCE Greece. Distress flares are still used by mariners in emergencies, while lighthouses and navigation lights are used to communicate navigation hazards.

The heliograph uses a mirror to reflect sunlight to a distant observer. When a signaler tilts the mirror to reflect sunlight, the distant observer sees flashes of light that can be used to transmit a prearranged signaling code. Naval ships often use signal lamps and Morse code in a similar way.

Aircraft pilots often use visual approach slope indicator (VASI) projected light systems to land safely, especially at night. Military aircraft landing on an aircraft carrier use a similar system to land correctly on a carrier deck. The coloured light system communicates the aircraft's height relative to a standard landing glideslope. As well, airport control towers still use Aldis lamps to transmit instructions to aircraft whose radios have failed.

In the present day a variety of electronic systems optically transmit and receive information carried by pulses of light. Fiber-optic communication cables are now employed to send the great majority of the electronic data and long distance telephone calls that are not conveyed by either radio, terrestrial microwave or satellite. Free-space optical communications are also used every day in various applications.

Semaphore Line

A replica of one of Chappe's semaphore towers (18th century).

A 'semaphore telegraph', also called a 'semaphore line', 'optical telegraph', 'shutter telegraph chain', 'Chappe telegraph', or 'Napoleonic semaphore', is a system used for conveying information by means of visual signals, using towers with pivoting arms or shutters, also known as blades or paddles. Information is encoded by the position of the mechanical elements; it is read when the shutter is in a fixed position.

Semaphore lines were a precursor of the electrical telegraph. They were far faster than post riders for conveying a message over long distances, but far more expensive and less private than the electrical telegraph lines which would later replace them. The maximum distance that a pair of semaphore telegraph stations can bridge is limited by geography, weather and the availability of light; thus, in practical use, most optical telegraphs used lines of relay stations to bridge longer distances. Each relay station would also require its complement of skilled operator-observers to convey messages back and forth across the line.

The modern design of semaphores was first foreseen by the British polymath Robert Hooke, who first gave a vivid and comprehensive outline of visual telegraphy in an 1684 submission to the Royal Society. His proposal (which was motivated by military concerns following the Battle of Vienna the preceding year) was not put into practice during his lifetime.

The first operational optical semaphore line arrived in 1792, created by the French engineer Claude Chappe and his brothers, who succeeded in covering France with a network of 556 stations stretching a total distance of 4,800 kilometres (3,000 mi). It was used for military and national communications until the 1850s.

Many national services adopted signaling systems different from the Chappe system. For example, Britain and Sweden adopted systems of shuttered panels (in contradiction to the Chappe brothers' contention that angled rods are more visible). In Spain, the engineer Agustín de Betancourt developed his own system which was adopted by that state. This system was considered by many experts in Europe better than Chappe's, even in France.

These systems were popular in the late 18th to early 19th century but could not compete with the electrical telegraph, and went completely out of service by 1880.

Semaphore Signal Flags

A naval signaler transmitting a message by flag semaphore (2002).

Semaphore Flags is the system for conveying information at a distance by means of visual signals with hand-held flags, rods, disks, paddles, or occasionally bare or gloved hands. Information is encoded by the position of the flags, objects or arms; it is read when they are in a fixed position.

Semaphores were adopted and widely used (with hand-held flags replacing the mechanical arms of shutter semaphores) in the maritime world in the 19th century.

They are still used during underway replenishment at sea and are acceptable for emergency communication in daylight or, using lighted wands instead of flags, at night.

The newer flag semaphore system uses two short poles with square flags, which a signaler holds in different positions to convey letters of the alphabet and numbers. The transmitter holds one pole in each hand, and extends each arm in one of eight possible directions. Except for in the rest position, the flags cannot overlap. The flags are colored differently based on whether the signals are sent by sea or by land. At sea, the flags are colored red and yellow (the Oscar flags), while on land, they are white and blue (the Papa flags). Flags are not required, they just make the characters more obvious.

Optical Fiber

Optical fiber is the most common type of channel for optical communications. The transmitters in optical fiber links are generally light-emitting diodes (LEDs) or laser diodes. Infrared light, rather than visible light is used more commonly, because optical fibers transmit infrared wavelengths with less attenuation and dispersion. The signal encoding is typically simple intensity modulation, although historically optical phase and frequency modulation have been demonstrated in the lab. The need for periodic signal regeneration was largely superseded by the introduction of the erbium-doped fiber amplifier, which extended link distances at significantly lower cost.

Signal Lamps

An air traffic controller holding a signal light gun that can be used to direct aircraft experiencing a radio failure (2007).

Signal lamps (such as Aldis lamps), are visual signaling devices for optical communication (typically using Morse code). Modern signal lamps are a focused lamp which can produce a pulse of light. In large versions this pulse is achieved by opening and closing shutters mounted in front of the lamp, either via a manually operated pressure switch or, in later versions, automatically.

With hand held lamps, a concave mirror is tilted by a trigger to focus the light into pulses. The lamps are usually equipped with some form of optical sight, and are most commonly deployed on naval vessels and also used in airport control towers with coded aviation light signals.

Aviation light signals are used in the case of a radio failure, an aircraft not equipped with a radio, or in the case of a hearing-impaired pilot. Air traffic controlers have long used signal light guns to direct such aircraft. The light gun's lamp has a focused bright beam capable of emitting three different colors: red, white and green. These colors may be flashing or steady, and provide different instructions to aircraft in flight or on the ground (for example, "cleared to land" or "cleared for takeoff"). Pilots can acknowledge the instructions by wiggling their plane's wings, moving their ailerons if they are on the ground, or by flashing their landing or navigation lights during night time. Only 12 simple standardized instructions are directed at aircraft using signal light guns as the system is not utilized with Morse code.

Photophone

The photophone (originally given an alternate name, radiophone) is a communication device which allowed for the transmission of speech on a beam of light. It was invented jointly by Alexander Graham Bell and his assistant Charles Sumner Tainter on February 19, 1880, at Bell's 1325 'L' Street laboratory in Washington, D.C. Both were later to become full associates in the Volta Laboratory Association, created and financed by Bell.

On June 21, 1880, Bell's assistant transmitted a wireless voice telephone message of considerable distance, from the roof of the Franklin School to the window of Bell's laboratory, some 213 meters (about 700 ft.) away.

Bell believed the photophone was his most important invention. Of the 18 patents granted in Bell's name alone, and the 12 he shared with his collaborators, four were for the photophone, which Bell referred to as his *'greatest achievement'*, telling a reporter shortly before his death that the photophone was *"the greatest invention [I have] ever made, greater than the telephone"*.

The photophone was a precursor to the fiber-optic communication systems which achieved popular worldwide usage starting in the 1980s. The master patent for the photophone (U.S. Patent 235,199 *Apparatus for Signalling and Communicating, called Photophone*), was issued in December 1880, many decades before its principles came to have practical applications.

Free-space Optical Communication

Free-space optics (FSO) systems are generally employed for 'last mile' telecommunications and can function over distances of several kilometers as long as there is a clear line of sight between the source and the destination, and the optical receiver can reliably decode the transmitted information. Other free-space systems can provide high-data-rate, long-range links using small, low-mass, low-power-consumption subsystems.

More generally, transmission of unguided optical signals is known as optical wireless communications (OWC). Examples include medium-range visible light communication and short-distance IrDA, using infrared LEDs.

Heliograph

Heliograph: Australians using a heliograph in North Africa (1940).

A heliograph is a wireless solar telegraph that signals by flashes of sunlight (generally using Morse code) reflected by a mirror. The flashes are produced by momentarily pivoting the mirror, or by interrupting the beam with a shutter.

The heliograph was a simple but effective instrument for instantaneous optical communication over long distances during the late 19th and early 20th century. Its main uses were in military, surveys and forest protection work. They were standard issue in the British and Australian armies until the 1960s, and were used by the Pakistani army as late as 1975.

Optical Fiber

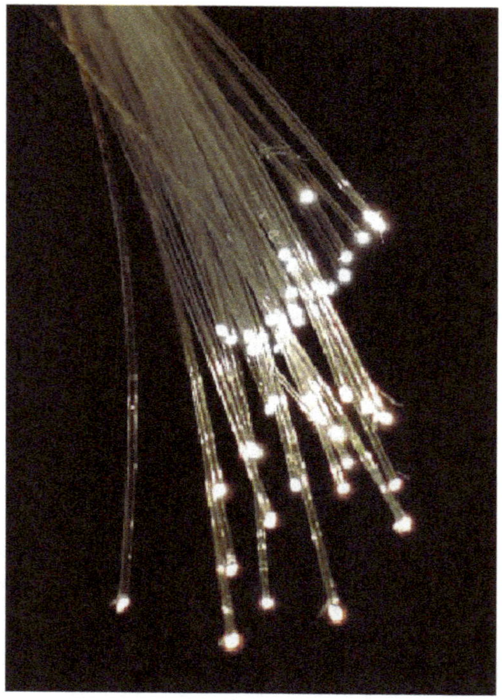

A bundle of optical fibers.

Fiber crew installing a 432-count fiber cable underneath the streets of Midtown Manhattan, New York City.

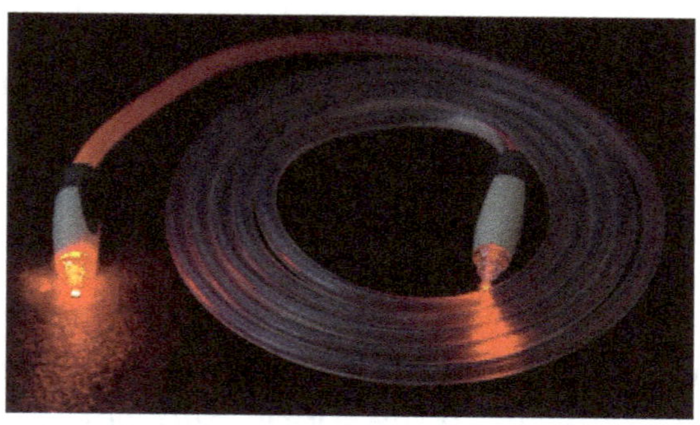

A TOSLINK fiber optic audio cable with red light being shone in one end transmits the light to the other end.

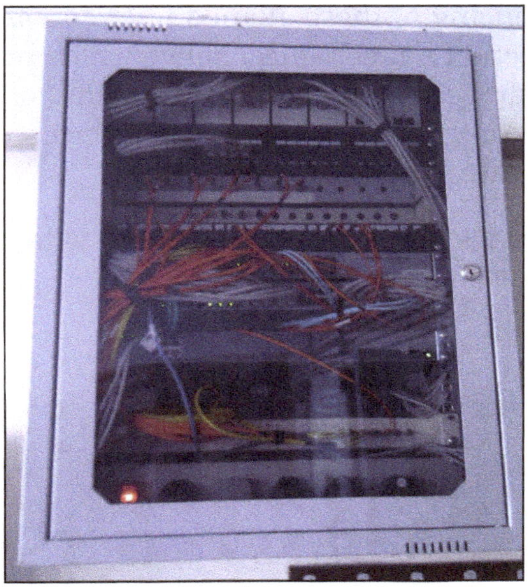

A wall-mount cabinet containing optical fiber interconnects. The yellow cables are single mode fibers; the orange and aqua cables are multi-mode fibers: 50/125 μm OM2 and 50/125 μm OM3 fibers respectively.

An optical fiber (or optical fibre) is a flexible, transparent fiber made by drawing glass (silica) or plastic to a diameter slightly thicker than that of a human hair. Optical fibers are used most often as a means to transmit light between the two ends of the fiber and find wide usage in fiber-optic communications, where they permit transmission over longer distances and at higher bandwidths (data rates) than wire cables. Fibers are used instead of metal wires because signals travel along them with lesser amounts of loss; in addition, fibers are also immune to electromagnetic interference, a problem from which metal wires suffer excessively. Fibers are also used for illumination, and are wrapped in bundles so that they may be used to carry images, thus allowing viewing in confined spaces, as in the case of a fiberscope. Specially designed fibers are also used for a variety of other applications, some of them being fiber optic sensors and fiber lasers.

Optical fibers typically include a transparent core surrounded by a transparent cladding material with a lower index of refraction. Light is kept in the core by the phenomenon of total internal reflection which causes the fiber to act as a waveguide. Fibers that support many propagation paths or transverse modes are called multi-mode fibers (MMF), while those that support a single mode are called single-mode fibers (SMF). Multi-mode fibers generally have a wider core diameter and are used for short-distance communication links and for applications where high power must be transmitted. Single-mode fibers are used for most communication links longer than 1,000 meters (3,300 ft).

An important aspect of a fiber optic communication is that of extension of the fiber optic cables such that the losses brought about by joining two different cables is kept to a minimum. Joining lengths of optical fiber often proves to be more complex than joining electrical wire or cable and involves careful cleaving of the fibers, perfect alignment of the fiber cores, and the splicing of these aligned fiber cores. For applications that demand a permanent connection a mechanical splice which holds the ends of the fibers together mechanically could be used or a fusion splice that uses heat to fuse the ends of the fibers together could be used. Temporary or semi-permanent connections are made by means of specialized optical fiber connectors.

The field of applied science and engineering concerned with the design and application of optical fibers is known as fiber optics.

History

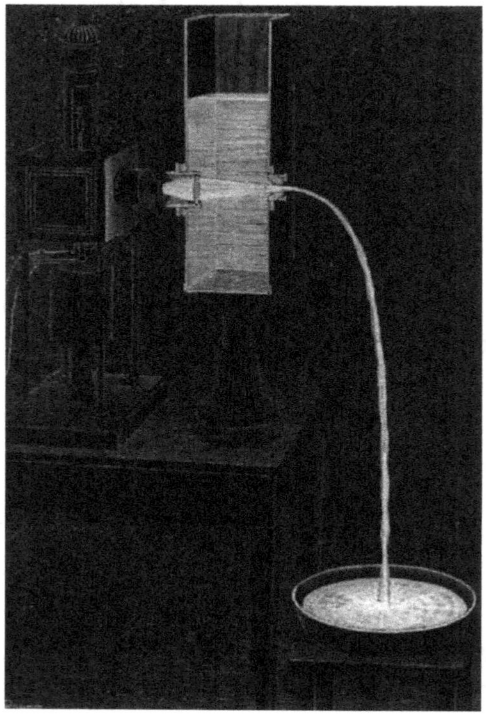

Daniel Colladon first described this "light fountain" or "light pipe" in an 1842 article titled On the reflections of a ray of light inside a parabolic liquid stream. This particular illustration comes from a later article by Colladon, in 1884.

Guiding of light by refraction, the principle that makes fiber optics possible, was first demonstrated by Daniel Colladon and Jacques Babinet in Paris in the early 1840s. John Tyndall included a demonstration of it in his public lectures in London, 12 years later. Tyndall also wrote about the property of total internal reflection in an introductory book about the nature of light in 1870:

When the light passes from air into water, the refracted ray is bent *towards* the perpendicular... When the ray passes from water to air it is bent *from* the perpendicular... If the angle which the ray in water encloses with the perpendicular to the surface be greater than 48 degrees, the ray will not quit the water at all: it will be *totally reflected* at the surface.... The angle which marks the limit where total reflection begins is called the limiting angle of the medium. For water this angle is 48°27′, for flint glass it is 38°41′, while for diamond it is 23°42′.

Unpigmented human hairs have also been shown to act as an optical fiber.

Practical applications, such as close internal illumination during dentistry, appeared early in the twentieth century. Image transmission through tubes was demonstrated independently by the radio experimenter Clarence Hansell and the television pioneer John Logie Baird in the 1920s. The principle was first used for internal medical examinations by Heinrich Lamm in the following decade. Modern optical fibers, where the glass fiber is coated with a transparent cladding to offer a more suitable refractive index, appeared later in the decade. Development then focused on fiber bundles for image transmission. Harold Hopkins and Narinder Singh Kapany at Imperial College in London achieved low-loss light transmission through a 75 cm long bundle which combined several thousand fibers. Their article titled "A flexible fibrescope, using static scanning" was published in the journal *Nature* in 1954. The first fiber optic semi-flexible gastroscope was patented by Basil Hirschowitz, C. Wilbur Peters, and Lawrence E. Curtiss, researchers at the University of Michigan, in 1956. In the process of developing the gastroscope, Curtiss produced the first glass-clad fibers; previous optical fibers had relied on air or impractical oils and waxes as the low-index cladding material.

A variety of other image transmission applications soon followed.

In 1880 Alexander Graham Bell and Sumner Tainter invented the Photophone at the Volta Laboratory in Washington, D.C., to transmit voice signals over an optical beam. It was an advanced form of telecommunications, but subject to atmospheric interferences and impractical until the secure transport of light that would be offered by fiber-optical systems. In the late 19th and early 20th centuries, light was guided through bent glass rods to illuminate body cavities. Jun-ichi Nishizawa, a Japanese scientist at Tohoku University, also proposed the use of optical fibers for communications in 1963, as stated in his book published in 2004 in India. Nishizawa invented other technologies that contributed to the development of optical fiber communications, such as the graded-index optical fiber as a channel for transmitting light from semiconductor lasers. The first working fiber-optical data transmission system was demonstrated by German physicist Manfred Börner at Telefunken Research Labs in Ulm in 1965, which was followed by the first patent application for this technology in 1966. NASA used fiber optics in the television cameras that were sent to the moon. At the time, the use in the cameras was classified *confidential*, and employees handling the cameras had to be supervised by someone with an appropriate security clearance.

Charles K. Kao and George A. Hockham of the British company Standard Telephones and Cables (STC) were the first to promote the idea that the attenuation in optical fibers could be reduced below 20 decibels per kilometer (dB/km), making fibers a practical communication medium. They proposed that the attenuation in fibers available at the time was caused by impurities that could be removed, rather than by fundamental physical effects such as scattering. They correctly and systematically theorized the light-loss properties for optical fiber, and pointed out the right material to use for such fibers — silica glass with high purity. This discovery earned Kao the Nobel Prize in Physics in 2009.

The crucial attenuation limit of 20 dB/km was first achieved in 1970, by researchers Robert D. Maurer, Donald Keck, Peter C. Schultz, and Frank Zimar working for American glass maker Corning Glass Works, now Corning Incorporated. They demonstrated a fiber with 17 dB/km attenuation by doping silica glass with titanium. A few years later they produced a fiber with only 4 dB/km attenuation using germanium dioxide as the core dopant. Such low attenuation ushered in the era of optical fiber telecommunication. In 1981, General Electric produced fused quartz ingots that could be drawn into strands 25 miles (40 km) long.

The Italian research center CSELT worked with Corning to develop practical optical fiber cables, resulting in the first metropolitan fiber optic cable being deployed in Torino in 1977. CSELT also developed an early technique for splicing optical fibers, called Springroove.

Attenuation in modern optical cables is far less than in electrical copper cables, leading to long-haul fiber connections with repeater distances of 70–150 kilometers (43–93 mi). The erbium-doped fiber amplifier, which reduced the cost of long-distance fiber systems by reducing or eliminating optical-electrical-optical repeaters, was co-developed by teams led by David N. Payne of the University of Southampton and Emmanuel Desurvire at Bell Labs in 1986. Robust modern optical fiber uses glass for both core and sheath, and is therefore less prone to aging. It was invented by Gerhard Bernsee of Schott Glass in Germany in 1973.

The emerging field of photonic crystals led to the development in 1991 of photonic-crystal fiber, which guides light by diffraction from a periodic structure, rather than by total internal reflection. The first photonic crystal fibers became commercially available in 2000. Photonic crystal fibers can carry higher power than conventional fibers and their wavelength-dependent properties can be manipulated to improve performance.

Uses

Communication

Optical fiber can be used as a medium for telecommunication and computer networking because it is flexible and can be bundled as cables. It is especially advantageous for long-distance communications, because light propagates through the fiber with little attenuation compared to electrical cables. This allows long distances to be spanned with few repeaters.

The per-channel light signals propagating in the fiber have been modulated at rates as high as 111 gigabits per second (Gbit/s) by NTT, although 10 or 40 Gbit/s is typical in deployed systems. In June 2013, researchers demonstrated transmission of 400 Gbit/s over a single channel using 4-mode orbital angular momentum multiplexing.

Each fiber can carry many independent channels, each using a different wavelength of light (wavelength-division multiplexing (WDM)). The net data rate (data rate without overhead bytes) per fiber is the per-channel data rate reduced by the FEC overhead, multiplied by the number of channels (usually up to eighty in commercial dense WDM systems as of 2008). As of 2011 the record for bandwidth on a single core was 101 Tbit/s (370 channels at 273 Gbit/s each). The record for a multi-core fiber as of January 2013 was 1.05 petabits per second. In 2009, Bell Labs broke the 100 (petabit per second)×kilometer barrier (15.5 Tbit/s over a single 7,000 km fiber).

For short distance application, such as a network in an office building, fiber-optic cabling can save space in cable ducts. This is because a single fiber can carry much more data than electrical cables such as standard category 5 Ethernet cabling, which typically runs at 100 Mbit/s or 1 Gbit/s speeds. Fiber is also immune to electrical interference; there is no cross-talk between signals in different cables, and no pickup of environmental noise. Non-armored fiber cables do not conduct electricity, which makes fiber a good solution for protecting communications equipment in high voltage environments, such as power generation facilities, or metal communication structures prone to lightning strikes. They can also be used in environments where explosive fumes are present, without danger of ignition. Wiretapping (in this case, fiber tapping) is more difficult compared to electrical connections, and there are concentric dual-core fibers that are said to be tap-proof.

Fibers are often also used for short-distance connections between devices. For example, most high-definition televisions offer a digital audio optical connection. This allows the streaming of audio over light, using the TOSLINK protocol.

Advantages Over Copper Wiring

The advantages of optical fiber communication with respect to copper wire systems are:

Broad bandwidth

> A single optical fiber can carry over 3,000,000 full-duplex voice calls or 90,000 TV channels.

Immunity to electromagnetic interference

> Light transmission through optical fibers is unaffected by other electromagnetic radiation nearby. The optical fiber is electrically non-conductive, so it does not act as an antenna to pick up electromagnetic signals. Information traveling

inside the optical fiber is immune to electromagnetic interference, even electromagnetic pulses generated by nuclear devices.

Low attenuation loss over long distances

Attenuation loss can be as low as 0.2 dB/km in optical fiber cables, allowing transmission over long distances without the need for repeaters.

Electrical insulator

Optical fibers do not conduct electricity, preventing problems with ground loops and conduction of lightning. Optical fibers can be strung on poles alongside high voltage power cables.

Material cost and theft prevention

Conventional cable systems use large amounts of copper. Global copper prices experienced a boom in the 2000s, and copper has been a target of metal theft.

Security of information passed down the cable

Copper can be tapped with very little chance of detection.

Sensors

Fibers have many uses in remote sensing. In some applications, the sensor is itself an optical fiber. In other cases, fiber is used to connect a non-fiberoptic sensor to a measurement system. Depending on the application, fiber may be used because of its small size, or the fact that no electrical power is needed at the remote location, or because many sensors can be multiplexed along the length of a fiber by using different wavelengths of light for each sensor, or by sensing the time delay as light passes along the fiber through each sensor. Time delay can be determined using a device such as an *optical time-domain reflectometer*.

Optical fibers can be used as sensors to measure strain, temperature, pressure and other quantities by modifying a fiber so that the property to measure modulates the intensity, phase, polarization, wavelength, or transit time of light in the fiber. Sensors that vary the intensity of light are the simplest, since only a simple source and detector are required. A particularly useful feature of such fiber optic sensors is that they can, if required, provide distributed sensing over distances of up to one meter. In contrast, highly localized measurements can be provided by integrating miniaturized sensing elements with the tip of the fiber. These can be implemented by various micro- and nanofabrication technologies, such that they do not exceed the microscopic boundary of the fiber tip, allowing such applications as insertion into blood vessels via hypodermic needle.

Extrinsic fiber optic sensors use an optical fiber cable, normally a multi-mode one, to transmit modulated light from either a non-fiber optical sensor—or an electronic sensor connected to an optical transmitter. A major benefit of extrinsic sensors is their ability to reach otherwise inaccessible places. An example is the measurement of temperature inside aircraft jet engines by using a fiber to transmit radiation into a radiation pyrometer outside the engine. Extrinsic sensors can be used in the same way to measure the internal temperature of electrical transformers, where the extreme electromagnetic fields present make other measurement techniques impossible. Extrinsic sensors measure vibration, rotation, displacement, velocity, acceleration, torque, and twisting. A solid state version of the gyroscope, using the interference of light, has been developed. The *fiber optic gyroscope (FOG)* has no moving parts, and exploits the *Sagnac effect* to detect mechanical rotation.

Common uses for fiber optic sensors includes advanced intrusion detection security systems. The light is transmitted along a fiber optic sensor cable placed on a fence, pipeline, or communication cabling, and the returned signal is monitored and analyzed for disturbances. This return signal is digitally processed to detect disturbances and trip an alarm if an intrusion has occurred.

Optical fibers are widely used as components of optical chemical sensors and optical biosensors.

Power Transmission

Optical fiber can be used to transmit power using a photovoltaic cell to convert the light into electricity. While this method of power transmission is not as efficient as conventional ones, it is especially useful in situations where it is desirable not to have a metallic conductor as in the case of use near MRI machines, which produce strong magnetic fields. Other examples are for powering electronics in high-powered antenna elements and measurement devices used in high-voltage transmission equipment.

Other Uses

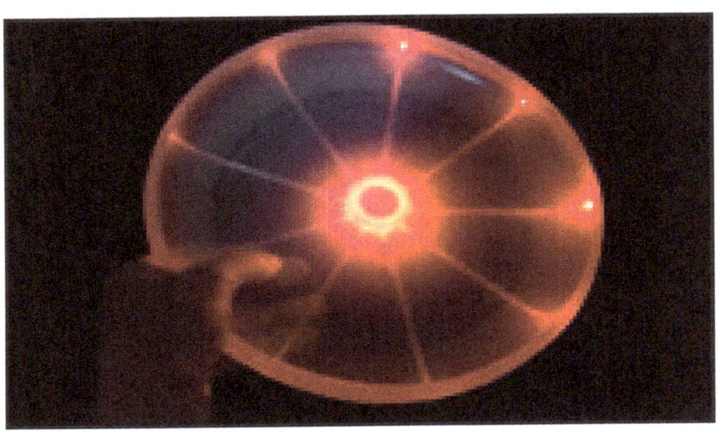

A frisbee illuminated by fiber optics.

Light reflected from optical fiber illuminates exhibited model.

Optical fibers have a wide number of applications. They are used as light guides in medical and other applications where bright light needs to be shone on a target without a clear line-of-sight path. In some buildings, optical fibers route sunlight from the roof to other parts of the building. Optical fiber lamps are used for illumination in decorative applications, including signs, art, toys and artificial Christmas trees. Swarovski boutiques use optical fibers to illuminate their crystal showcases from many different angles while only employing one light source. Optical fiber is an intrinsic part of the light-transmitting concrete building product, LiTraCon.

Use of optical fiber in a decorative lamp or nightlight.

Optical fiber is also used in imaging optics. A coherent bundle of fibers is used, sometimes along with lenses, for a long, thin imaging device called an endoscope, which is used to view objects through a small hole. Medical endoscopes are used for minimally invasive exploratory or surgical procedures. Industrial endoscopes are used for inspecting anything hard to reach, such as jet engine interiors. Many microscopes use fiber-optic light sources to provide intense illumination of samples being studied.

In spectroscopy, optical fiber bundles transmit light from a spectrometer to a substance that cannot be placed inside the spectrometer itself, in order to analyze its composition. A spectrometer analyzes substances by bouncing light off and through them. By using fibers, a spectrometer can be used to study objects remotely.

An optical fiber doped with certain rare earth elements such as erbium can be used as the gain medium of a laser or optical amplifier. Rare-earth-doped optical fibers can be used to provide signal amplification by splicing a short section of doped fiber into a regular (undoped) optical fiber line. The doped fiber is optically pumped with a second laser wavelength that is coupled into the line in addition to the signal wave. Both wavelengths of light are transmitted through the doped fiber, which transfers energy from the second pump wavelength to the signal wave. The process that causes the amplification is stimulated emission.

Optical fiber is also widely exploited as a nonlinear medium. The glass medium supports a host of nonlinear optical interactions, and the long interaction lengths possible in fiber facilitate a variety of phenomena, which are harnessed for applications and fundamental investigation. Conversely, fiber nonlinearity can have deleterious effects on optical signals, and measures are often required to minimize such unwanted effects.

Optical fibers doped with a wavelength shifter collect scintillation light in physics experiments.

Fiber optic sights for handguns, rifles, and shotguns use pieces of optical fiber to improve visibility of markings on the sight.

Principle of Operation

An overview of the operating principles of the optical fiber.

An optical fiber is a cylindrical dielectric waveguide (nonconducting waveguide) that transmits light along its axis, by the process of total internal reflection. The fiber consists of a *core* surrounded by a cladding layer, both of which are made of dielectric materials. To confine the optical signal in the core, the refractive index of the core must be greater than that of the cladding. The boundary between the core and cladding may either be abrupt, in *step-index fiber*, or gradual, in *graded-index fiber*.

Index of Refraction

The index of refraction (or refractive index) is a way of measuring the speed of light in a material. Light travels fastest in a vacuum, such as in outer space. The speed of light in a vacuum is about 300,000 kilometers (186,000 miles) per second. The refractive index of a medium is calculated by dividing the speed of light in a vacuum by the speed of light in that medium. The refractive index of a vacuum is therefore 1, by definition. A typical singlemode fiber used for telecommunications has a cladding made of pure silica, with an index of 1.444 at 1,500 nm, and a core of doped silica with an index around 1.4475. The larger the index of refraction, the slower light travels in that medium. From this information, a simple rule of thumb is that a signal using optical fiber for communication will travel at around 200,000 kilometers per second. To put it another way, the signal will take 5 milliseconds to travel 1,000 kilometers in fiber. Thus a phone call carried by fiber between Sydney and New York, a 16,000-kilometer distance, means that there is a minimum delay of 80 milliseconds (about of a second) between when one caller speaks and the other hears. (The fiber in this case will probably travel a longer route, and there will be additional delays due to communication equipment switching and the process of encoding and decoding the voice onto the fiber).

Total Internal Reflection

When light traveling in an optically dense medium hits a boundary at a steep angle (larger than the critical angle for the boundary), the light is completely reflected. This is called total internal reflection. This effect is used in optical fibers to confine light in the core. Light travels through the fiber core, bouncing back and forth off the boundary between the core and cladding. Because the light must strike the boundary with an angle greater than the critical angle, only light that enters the fiber within a certain range of angles can travel down the fiber without leaking out. This range of angles is called the acceptance cone of the fiber. The size of this acceptance cone is a function of the refractive index difference between the fiber's core and cladding.

In simpler terms, there is a maximum angle from the fiber axis at which light may enter the fiber so that it will propagate, or travel, in the core of the fiber. The sine of this maximum angle is the numerical aperture (NA) of the fiber. Fiber with a larger NA requires less precision to splice and work with than fiber with a smaller NA. Single-mode fiber has a small NA.

Multi-mode Fiber

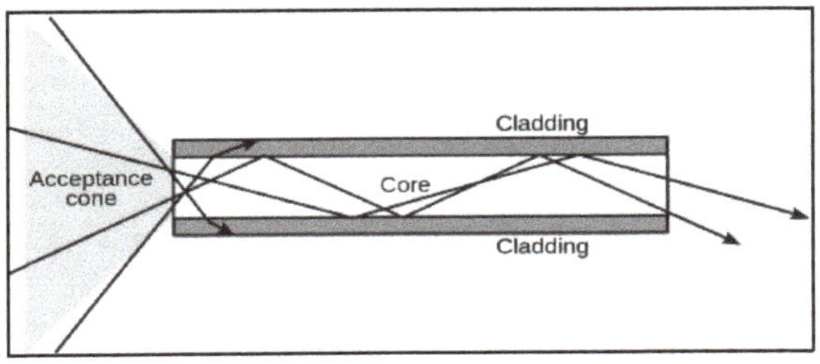

The propagation of light through a multi-mode optical fiber.

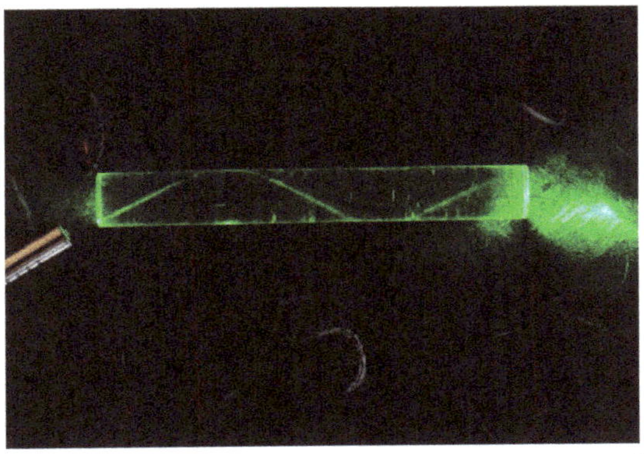

A laser bouncing down an acrylic rod, illustrating the total internal reflection of light in a multi-mode optical fiber.

Fiber with large core diameter (greater than 10 micrometers) may be analyzed by geometrical optics. Such fiber is called *multi-mode fiber*, from the electromagnetic analysis. In a step-index multi-mode fiber, rays of light are guided along the fiber core by total internal reflection. Rays that meet the core-cladding boundary at a high angle (measured relative to a line normal to the boundary), greater than the critical angle for this boundary, are completely reflected. The critical angle (minimum angle for total internal reflection) is determined by the difference in index of refraction between the core and cladding materials. Rays that meet the boundary at a low angle are refracted from the core into the cladding, and do not convey light and hence information along the fiber. The critical angle determines the acceptance angle of the fiber, often reported as a numerical aperture. A high numerical aperture allows light to propagate down the fiber in rays both close to the axis and at various angles, allowing efficient coupling of light into the fiber. However, this high numerical aperture increases the amount of dispersion as rays at different angles have different path lengths and therefore take different times to traverse the fiber.

In graded-index fiber, the index of refraction in the core decreases continuously between the axis and the cladding. This causes light rays to bend smoothly as they approach the cladding, rather than reflecting abruptly from the core-cladding boundary. The resulting curved paths reduce multi-path dispersion because high angle rays pass more through the lower-index periphery of the core, rather than the high-index center.

The index profile is chosen to minimize the difference in axial propagation speeds of the various rays in the fiber. This ideal index profile is very close to a parabolic relationship between the index and the distance from the axis.

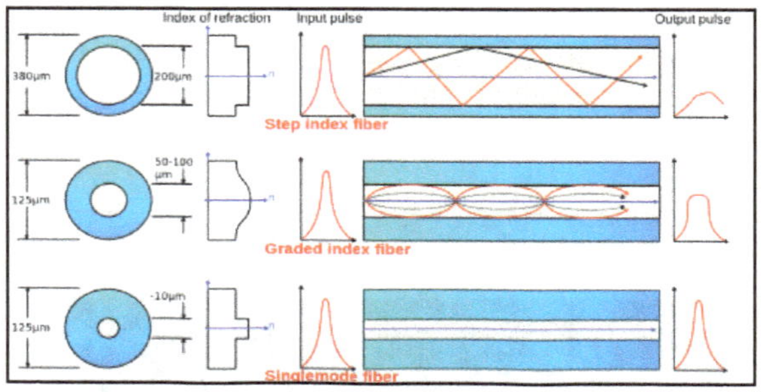

Optical fiber types.

Single-mode Fiber

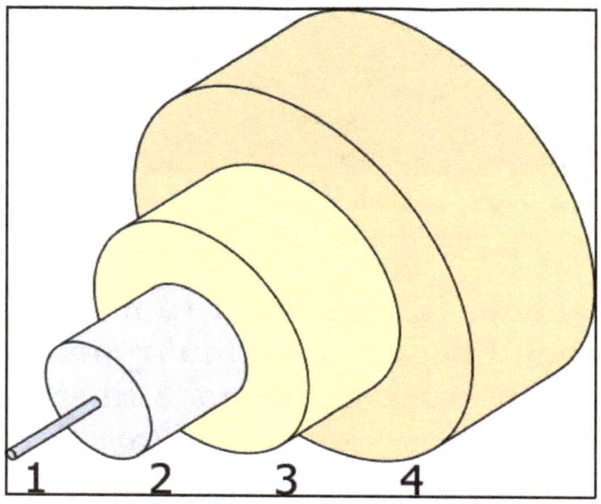

The structure of a typical single-mode fiber.
1. Core: 8 μm diameter.
2. Cladding: 125 μm dia.
3. Buffer: 250 μm dia.
4. Jacket: 400 μm dia.

Fiber with a core diameter less than about ten times the wavelength of the propagating light cannot be modeled using geometric optics. Instead, it must be analyzed as an electromagnetic structure, by solution of Maxwell's equations as reduced to the electromagnetic wave equation. The electromagnetic analysis may also be required to understand behaviors such as speckle that occur when coherent light propagates in multi-mode fiber. As an optical waveguide, the fiber supports one or more confined transverse modes by which light can propagate along the fiber. Fiber supporting only one mode is called *single-mode* or *mono-mode fiber*. The behavior of larger-core multi-mode fiber can also be modeled using the wave equation, which shows that such fiber supports more than one mode of propagation (hence the name). The results of such modeling of multi-mode fiber approximately agree with the predictions of geometric optics, if the fiber core is large enough to support more than a few modes.

The waveguide analysis shows that the light energy in the fiber is not completely confined in the core. Instead, especially in single-mode fibers, a significant fraction of the energy in the bound mode travels in the cladding as an evanescent wave.

The most common type of single-mode fiber has a core diameter of 8–10 micrometers and is designed for use in the near infrared. The mode structure depends on the wavelength of the light used, so that this fiber actually supports a small number of additional modes at visible wavelengths. Multi-mode fiber, by comparison, is manufactured with core diameters as small as 50 micrometers and as large as hundreds of micrometers. The normalized frequency V for this fiber should be less than the first zero of the Bessel function J_0 (approximately 2.405).

Special-purpose Fiber

Some special-purpose optical fiber is constructed with a non-cylindrical core and/or cladding layer, usually with an elliptical or rectangular cross-section. These include polarization-maintaining fiber and fiber designed to suppress whispering gallery mode propagation. Polarization-maintaining fiber is a unique type of fiber that is commonly used in fiber optic sensors due to its ability to maintain the polarization of the light inserted into it.

Photonic-crystal fiber is made with a regular pattern of index variation (often in the form of cylindrical holes that run along the length of the fiber). Such fiber uses diffraction effects instead of or in addition to total internal reflection, to confine light to the fiber's core. The properties of the fiber can be tailored to a wide variety of applications.

Mechanisms of Attenuation

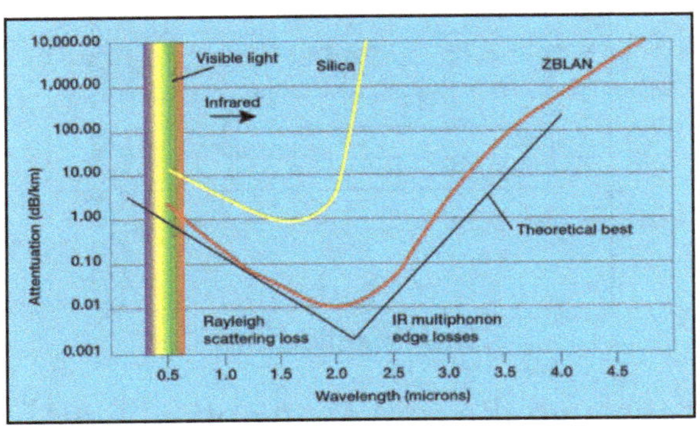

Light attenuation by ZBLAN and silica fibers.

Attenuation in fiber optics, also known as transmission loss, is the reduction in intensity of the light beam (or signal) as it travels through the transmission medium. Attenuation coefficients in fiber optics usually use units of dB/km through the medium due to the relatively high quality of transparency of modern optical transmission media. The medium is usually a fiber of silica glass that confines the incident light beam to the inside. Attenuation is an important factor limiting the transmission of a digital signal across large distances.

Thus, much research has gone into both limiting the attenuation and maximizing the amplification of the optical signal. Empirical research has shown that attenuation in optical fiber is caused primarily by both scattering and absorption. Single-mode optical fibers can be made with extremely low loss. Corning's SMF-28 fiber, a standard single-mode fiber for telecommunications wavelengths, has a loss of 0.17 dB/km at 1550 nm. For example, an 8 km length of SMF-28 transmits nearly 75% of light at 1,550 nm. It has been noted that if ocean water was as clear as fiber, one could see all the way to the bottom even of the Marianas Trench in the Pacific Ocean, a depth of 36,000 feet.

Light Scattering

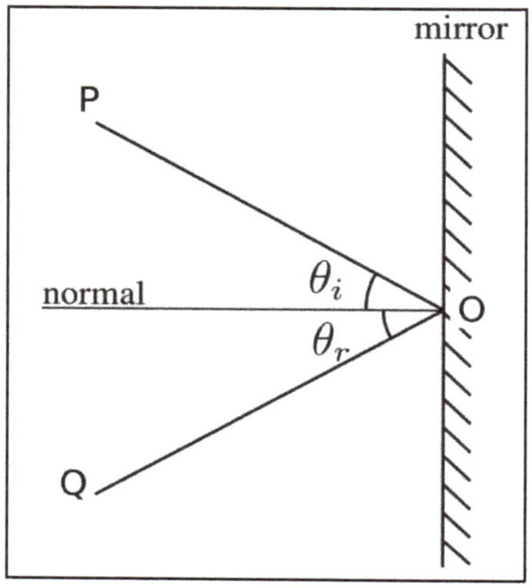

Specular reflection.

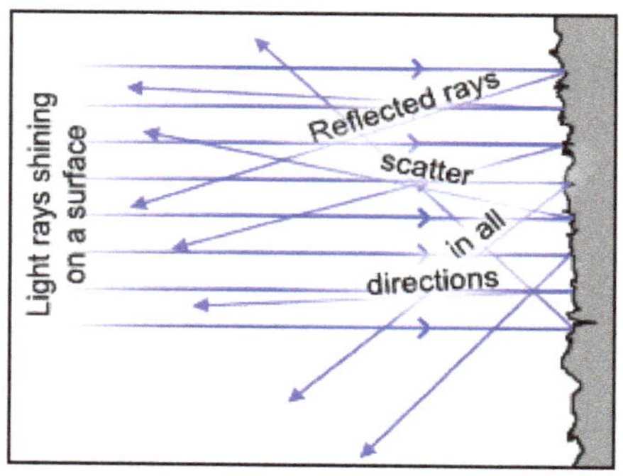

Diffuse reflection.

The propagation of light through the core of an optical fiber is based on total internal reflection of the lightwave. Rough and irregular surfaces, even at the molecular level, can cause light rays to be reflected in random directions. This is called diffuse reflection or scattering, and it is typically characterized by wide variety of reflection angles.

Light scattering depends on the wavelength of the light being scattered. Thus, limits to spatial scales of visibility arise, depending on the frequency of the incident light-wave and the physical dimension (or spatial scale) of the scattering center, which is typically in the form of some specific micro-structural feature. Since visible light has a wavelength of the order of one micrometer (one millionth of a meter) scattering centers will have dimensions on a similar spatial scale.

Thus, attenuation results from the incoherent scattering of light at internal surfaces and interfaces. In (poly)crystalline materials such as metals and ceramics, in addition to pores, most of the internal surfaces or interfaces are in the form of grain boundaries that separate tiny regions of crystalline order. It has recently been shown that when the size of the scattering center (or grain boundary) is reduced below the size of the wavelength of the light being scattered, the scattering no longer occurs to any significant extent. This phenomenon has given rise to the production of transparent ceramic materials.

Similarly, the scattering of light in optical quality glass fiber is caused by molecular level irregularities (compositional fluctuations) in the glass structure. Indeed, one emerging school of thought is that a glass is simply the limiting case of a polycrystalline solid. Within this framework, "domains" exhibiting various degrees of short-range order become the building blocks of both metals and alloys, as well as glasses and ceramics. Distributed both between and within these domains are micro-structural defects that provide the most ideal locations for light scattering. This same phenomenon is seen as one of the limiting factors in the transparency of IR missile domes.

At high optical powers, scattering can also be caused by nonlinear optical processes in the fiber.

UV-Vis-IR Absorption

In addition to light scattering, attenuation or signal loss can also occur due to selective absorption of specific wavelengths, in a manner similar to that responsible for the appearance of color. Primary material considerations include both electrons and molecules as follows:

- At the electronic level, it depends on whether the electron orbitals are spaced (or "quantized") such that they can absorb a quantum of light (or photon) of a specific wavelength or frequency in the ultraviolet (UV) or visible ranges. This is what gives rise to color.
- At the atomic or molecular level, it depends on the frequencies of atomic or molecular vibrations or chemical bonds, how close-packed its atoms or molecules are, and whether or not the atoms or molecules exhibit long-range order. These factors will determine the capacity of the material transmitting longer wavelengths in the infrared (IR), far IR, radio and microwave ranges.

The design of any optically transparent device requires the selection of materials based upon knowledge of its properties and limitations. The Lattice absorption characteristics observed at the lower frequency regions (mid IR to far-infrared wavelength range) define the long-wavelength transparency limit of the material. They are the result of the interactive coupling between the motions of thermally induced vibrations of the constituent atoms and molecules of the solid lattice and the incident light wave radiation. Hence, all materials are bounded by limiting regions of absorption caused by atomic and molecular vibrations (bond-stretching)in the far-infrared (>10 μm).

Thus, multi-phonon absorption occurs when two or more phonons simultaneously interact to produce electric dipole moments with which the incident radiation may couple. These dipoles can absorb energy from the incident radiation, reaching a maximum coupling with the radiation when the frequency is equal to the fundamental vibrational mode of the molecular dipole (e.g. Si-O bond) in the far-infrared, or one of its harmonics.

The selective absorption of infrared (IR) light by a particular material occurs because the selected frequency of the light wave matches the frequency (or an integer multiple of the frequency) at which the particles of that material vibrate. Since different atoms and molecules have different natural frequencies of vibration, they will selectively absorb different frequencies (or portions of the spectrum) of infrared (IR) light.

Reflection and transmission of light waves occur because the frequencies of the light waves do not match the natural resonant frequencies of vibration of the objects. When IR light of these frequencies strikes an object, the energy is either reflected or transmitted.

Manufacturing

Materials

Glass optical fibers are almost always made from silica, but some other materials, such as fluorozirconate, fluoroaluminate, and chalcogenide glasses as well as crystalline materials like sapphire, are used for longer-wavelength infrared or other specialized applications. Silica and fluoride glasses usually have refractive indices of about 1.5, but some materials such as the chalcogenides can have indices as high as 3. Typically the index difference between core and cladding is less than one percent.

Plastic optical fibers (POF) are commonly step-index multi-mode fibers with a core diameter of 0.5 millimeters or larger. POF typically have higher attenuation coefficients than glass fibers, 1 dB/m or higher, and this high attenuation limits the range of POF-based systems.

Silica

Silica exhibits fairly good optical transmission over a wide range of wavelengths. In the near-infrared (near IR) portion of the spectrum, particularly around 1.5 μm, silica can have extremely low absorption and scattering losses of the order of 0.2 dB/km. Such remarkably low losses are possible only because ultra-pure silicon is available, it being essential for manufacturing integrated circuits and discrete transistors. A high transparency in the 1.4-μm region is achieved by maintaining a low concentration of hydroxyl groups (OH). Alternatively, a high OH concentration is better for transmission in the ultraviolet (UV) region.

Silica can be drawn into fibers at reasonably high temperatures, and has a fairly broad glass transformation range. One other advantage is that fusion splicing and cleaving of silica fibers is relatively effective. Silica fiber also has high mechanical strength against both pulling and even bending, provided that the fiber is not too thick and that the surfaces have been well prepared during processing. Even simple cleaving (breaking) of the ends of the fiber can provide nicely flat surfaces with acceptable optical quality. Silica is also relatively chemically inert. In particular, it is not hygroscopic (does not absorb water).

Silica glass can be doped with various materials. One purpose of doping is to raise the refractive index (e.g. with germanium dioxide (GeO_2) or aluminium oxide (Al_2O_3)) or to lower it (e.g. with fluorine or boron trioxide (B_2O_3)). Doping is also possible with laser-active ions (for example, rare earth-doped fibers) in order to obtain active fibers to be used, for example, in fiber amplifiers or laser applications. Both the fiber core and cladding are typically doped, so that the entire assembly (core and cladding) is effectively the same compound (e.g. an aluminosilicate, germanosilicate, phosphosilicate or borosilicate glass).

Particularly for active fibers, pure silica is usually not a very suitable host glass, because it exhibits a low solubility for rare earth ions. This can lead to quenching effects due to clustering of dopant ions. Aluminosilicates are much more effective in this respect.

Silica fiber also exhibits a high threshold for optical damage. This property ensures a low tendency for laser-induced breakdown. This is important for fiber amplifiers when utilized for the amplification of short pulses.

Because of these properties silica fibers are the material of choice in many optical applications, such as communications (except for very short distances with plastic optical fiber), fiber lasers, fiber amplifiers, and fiber-optic sensors. Large efforts put forth in the development of various types of silica fibers have further increased the performance of such fibers over other materials.

Fluoride Glass

Fluoride glass is a class of non-oxide optical quality glasses composed of fluorides of various metals. Because of their low viscosity, it is very difficult to completely avoid crystallization while processing it through the glass transition (or drawing the fiber from the melt). Thus, although heavy metal fluoride glasses (HMFG) exhibit very low optical attenuation, they are not only difficult to manufacture, but are quite fragile, and have poor resistance to moisture and other environmental attacks. Their best attribute is that they lack the absorption band associated with the hydroxyl (OH) group (3,200–3,600 cm^{-1}; i.e., 2,777–3,125 nm or 2.78–3.13 μm), which is present in nearly all oxide-based glasses.

An example of a heavy metal fluoride glass is the ZBLAN glass group, composed of zirconium, barium, lanthanum, aluminium, and sodium fluorides. Their main technological application is as optical waveguides in both planar and fiber form. They are advantageous especially in the mid-infrared (2,000–5,000 nm) range.

HMFGs were initially slated for optical fiber applications, because the intrinsic losses of a mid-IR fiber could in principle be lower than those of silica fibers, which are transparent only up to about 2 μm. However, such low losses were never realized in practice, and the fragility and high cost of fluoride fibers made them less than ideal as primary candidates. Later, the utility of fluoride fibers for various other applications was discovered. These include mid-IR spectroscopy, fiber optic sensors, thermometry, and imaging. Also, fluoride fibers can be used for guided lightwave transmission in media such as YAG (yttrium aluminium garnet) lasers at 2.9 μm, as required for medical applications (e.g. ophthalmology and dentistry).

Phosphate Glass

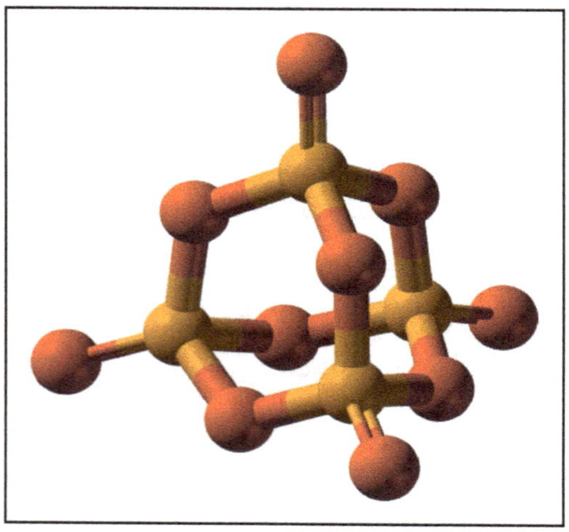

The P4O10 cagelike structure—the basic building block for phosphate glass.

Phosphate glass constitutes a class of optical glasses composed of metaphosphates of various metals. Instead of the SiO_4 tetrahedra observed in silicate glasses, the building block for this glass former is phosphorus pentoxide (P_2O_5), which crystallizes in at least four different forms. The most familiar polymorph comprises molecules of P_4O_{10}.

Phosphate glasses can be advantageous over silica glasses for optical fibers with a high concentration of doping rare earth ions. A mix of fluoride glass and phosphate glass is fluorophosphate glass.

Chalcogenide Glass

The chalcogens—the elements in group 16 of the periodic table—particularly sulfur (S), selenium (Se) and tellurium (Te)—react with more electropositive elements, such as silver, to form chalcogenides. These are extremely versatile compounds, in that they can be crystalline or amorphous, metallic or semiconducting, and conductors of ions or electrons. Glass containing chalcogenides can be used to make fibers for far infrared transmission.

Process

Preform

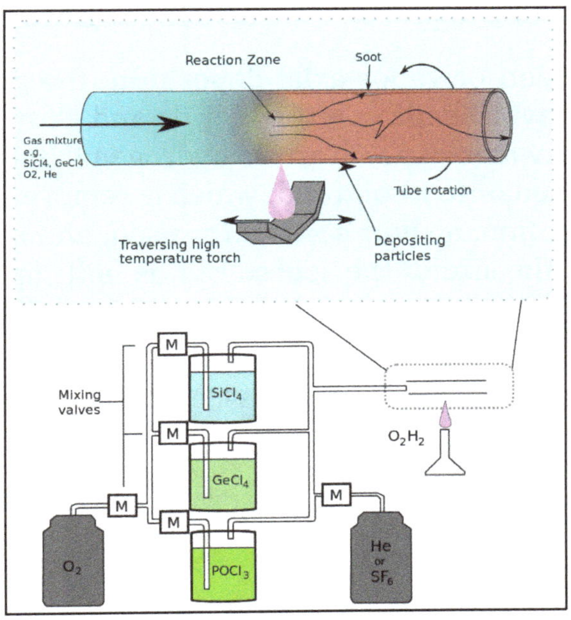

Illustration of the modified chemical vapor deposition (inside) process.

Standard optical fibers are made by first constructing a large-diameter "preform" with a carefully controlled refractive index profile, and then "pulling" the preform to form the long, thin optical fiber. The preform is commonly made by three chemical vapor deposition methods: *inside vapor deposition*, *outside vapor deposition*, and *vapor axial deposition.*

With *inside vapor deposition*, the preform starts as a hollow glass tube approximately 40 centimeters (16 in) long, which is placed horizontally and rotated slowly on a lathe. Gases such as silicon tetrachloride ($SiCl_4$) or germanium tetrachloride ($GeCl_4$) are injected with oxygen in the end of the tube. The gases are then heated by means of an external hydrogen burner, bringing the temperature of the gas up to 1,900 K (1,600 °C, 3,000 °F), where the tetrachlorides react with oxygen to produce silica or germania (germanium dioxide) particles. When the reaction conditions are chosen to allow this reaction to occur in the gas phase throughout the tube volume, in contrast to earlier techniques where the reaction occurred only on the glass surface, this technique is called *modified chemical vapor deposition (MCVD).*

The oxide particles then agglomerate to form large particle chains, which subsequently deposit on the walls of the tube as soot. The deposition is due to the large difference in temperature between the gas core and the wall causing the gas to push the particles outwards (this is known as thermophoresis). The torch is then traversed up and down the length of the tube to deposit the material evenly. After the torch has reached the end of the tube, it is then brought back to the beginning of the tube and the deposited particles are then melted to form a solid layer. This process is repeated until a sufficient amount of material has been deposited. For each layer the composition can be modified by varying the gas composition, resulting in precise control of the finished fiber's optical properties.

In outside vapor deposition or vapor axial deposition, the glass is formed by *flame hydrolysis*, a reaction in which silicon tetrachloride and germanium tetrachloride are oxidized by reaction with water (H_2O) in an oxyhydrogen flame. In outside vapor deposition the glass is deposited onto a solid rod, which is removed before further processing. In vapor axial deposition, a short *seed rod* is used, and a porous preform, whose length is not limited by the size of the source rod, is built up on its end. The porous preform is consolidated into a transparent, solid preform by heating to about 1,800 K (1,500 °C, 2,800 °F).

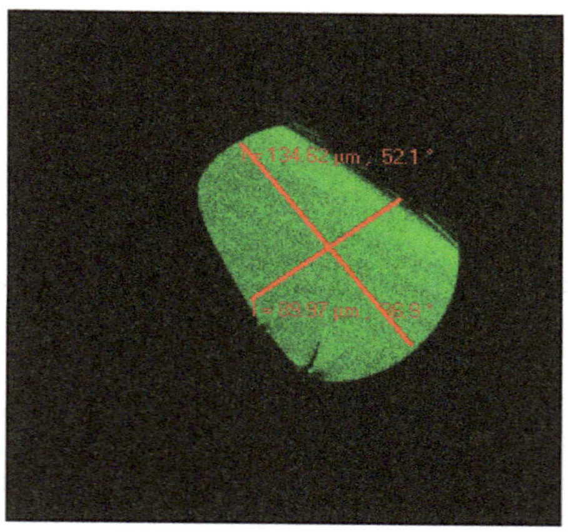

Cross-section of a fiber drawn from a D-shaped preform.

Typical communications fiber uses a circular preform. For some applications such as double-clad fibers another form is preferred. In fiber lasers based on double-clad fiber, an asymmetric shape improves the filling factor for laser pumping.

Because of the surface tension, the shape is smoothed during the drawing process, and the shape of the resulting fiber does not reproduce the sharp edges of the preform. Nevertheless, careful polishing of the preform is important, since any defects of the preform surface affect the optical and mechanical properties of the resulting fiber. In particular, the preform for the test-fiber shown in the figure was not polished well, and cracks are seen with the confocal optical microscope.

Drawing

The preform, however constructed, is placed in a device known as a drawing tower, where the preform tip is heated and the optical fiber is pulled out as a string. By measuring the resultant fiber width, the tension on the fiber can be controlled to maintain the fiber thickness.

Coatings

The light is guided down the core of the fiber by an optical cladding with a lower refractive index that traps light in the core through total internal reflection.

The cladding is coated by a buffer that protects it from moisture and physical damage. The buffer coating is what gets stripped off the fiber for termination or splicing. These coatings are UV-cured urethane acrylate composite or polyimide materials applied to the outside of the fiber during the drawing process. The coatings protect the very delicate strands of glass fiber—about the size of a human hair—and allow it to survive the rigors of manufacturing, proof testing, cabling and installation.

Today's glass optical fiber draw processes employ a dual-layer coating approach. An inner primary coating is designed to act as a shock absorber to minimize attenuation caused by microbending. An outer secondary coating protects the primary coating against mechanical damage and acts as a barrier to lateral forces. Sometimes a metallic armor layer is added to provide extra protection.

These fiber optic coating layers are applied during the fiber draw, at speeds approaching 100 kilometers per hour (60 mph). Fiber optic coatings are applied using one of two methods: *wet-on-dry* and *wet-on-wet*. In wet-on-dry, the fiber passes through a primary coating application, which is then UV cured—then through the secondary coating application, which is subsequently cured. In wet-on-wet, the fiber passes through both the primary and secondary coating applications, then goes to UV curing.

Fiber optic coatings are applied in concentric layers to prevent damage to the fiber during the drawing application and to maximize fiber strength and microbend resistance. Unevenly coated fiber will experience non-uniform forces when the coating expands or contracts, and is susceptible to greater signal attenuation. Under proper drawing and coating processes, the coatings are concentric around the fiber, continuous over the length of the application and have constant thickness.

Fiber optic coatings protect the glass fibers from scratches that could lead to strength degradation. The combination of moisture and scratches accelerates the aging and deterioration of fiber strength. When fiber is subjected to low stresses over a long period, fiber fatigue can occur. Over time or in extreme conditions, these factors combine to cause microscopic flaws in the glass fiber to propagate, which can ultimately result in fiber failure.

Three key characteristics of fiber optic waveguides can be affected by environmental conditions: strength, attenuation and resistance to losses caused by microbending. External fiber optic coatings protect glass optical fiber from environmental conditions that can affect the fiber's performance and long-term durability. On the inside, coatings ensure the reliability of the signal being carried and help minimize attenuation due to microbending.

Practical Issues

Cable Construction

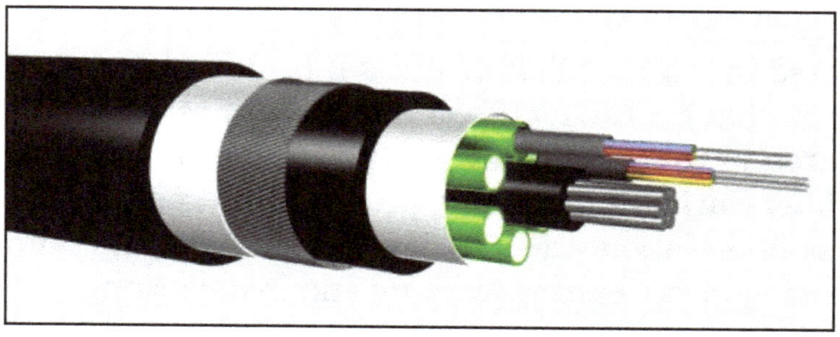

An optical fiber cable.

In practical fibers, the cladding is usually coated with a tough resin coating and an additional *buffer* layer, which may be further surrounded by a *jacket* layer, usually plastic. These layers add strength to the fiber but do not contribute to its optical wave guide properties. Rigid fiber assemblies sometimes put light-absorbing ("dark") glass between the fibers, to prevent light that leaks out of one fiber from entering another. This reduces cross-talk between the fibers, or reduces flare in fiber bundle imaging applications.

Modern cables come in a wide variety of sheathings and armor, designed for applications such as direct burial in trenches, high voltage isolation, dual use as power lines,installation in conduit, lashing to aerial telephone poles, submarine installation, and insertion in paved streets. Multi-fiber cable usually uses colored coatings and/or buffers to identify each strand. The cost of small fiber-count pole-mounted cables has greatly decreased due to the high demand for fiber to the home (FTTH) installations in Japan and South Korea.

Fiber cable can be very flexible, but traditional fiber's loss increases greatly if the fiber is bent with a radius smaller than around 30 mm. This creates a problem when the cable is bent around corners or wound around a spool, making FTTX installations more complicated. "Bendable fibers", targeted towards easier installation in home environments, have been standardized as ITU-T G.657. This type of fiber can be bent with a radius as low as 7.5 mm without adverse impact. Even more bendable fibers have been developed. Bendable fiber may also be resistant to fiber hacking, in which the signal in a fiber is surreptitiously monitored by bending the fiber and detecting the leakage.

Another important feature of cable is cable's ability to withstand horizontally applied force. It is technically called max tensile strength defining how much force can be applied to the cable during the installation period.

Some fiber optic cable versions are reinforced with aramid yarns or glass yarns as intermediary strength member. In commercial terms, usage of the glass yarns are more cost effective while no loss in mechanical durability of the cable. Glass yarns also protect the cable core against rodents and termites.

Termination and Splicing

ST connectors on multi-mode fiber.

Optical fibers are connected to terminal equipment by optical fiber connectors. These connectors are usually of a standard type such as *FC, SC, ST, LC, MTRJ*, or *SMA*. Optical fibers may be connected to each other by connectors or by *splicing*, that is, joining two fibers together to form a continuous optical waveguide. The generally accepted splicing method is arc fusion splicing, which melts the fiber ends together with an electric arc. For quicker fastening jobs, a "mechanical splice" is used.

Fusion splicing is done with a specialized instrument. The fiber ends are first stripped of their protective polymer coating (as well as the more sturdy outer jacket, if present). The ends are *cleaved* (cut) with a precision cleaver to make them perpendicular, and are placed into special holders in the fusion splicer. The splice is usually inspected via a magnified viewing screen to check the cleaves before and after the splice. The splicer uses small motors to align the end faces together, and emits a small spark between electrodes at the gap to burn off dust and moisture. Then the splicer generates a larger spark that raises the temperature above the melting point of the glass, fusing the ends together permanently. The location and energy of the spark is carefully controlled so that the molten core and cladding do not mix, and this minimizes optical loss. A splice loss estimate is measured by the splicer, by directing light through the cladding on one side and measuring the light leaking from the cladding on the other side. A splice loss under 0.1 dB is typical. The complexity of this process makes fiber splicing much more difficult than splicing copper wire.

Mechanical fiber splices are designed to be quicker and easier to install, but there is still the need for stripping, careful cleaning and precision cleaving. The fiber ends are aligned and held together by a precision-made sleeve, often using a clear index-matching gel that enhances the transmission of light across the joint. Such joints typically have higher optical loss and are less robust than fusion splices, especially if the gel is used. All splicing techniques involve installing an enclosure that protects the splice.

Fibers are terminated in connectors that hold the fiber end precisely and securely. A fiber-optic connector is basically a rigid cylindrical barrel surrounded by a sleeve that holds the barrel in its mating socket. The mating mechanism can be *push and click*, *turn and latch* (*bayonet mount*), or *screw-in* (*threaded*). The barrel is typically free to move within the sleeve, and may have a key that prevents the barrel and fiber from rotating as the connectors are mated.

A typical connector is installed by preparing the fiber end and inserting it into the rear of the connector body. Quick-set adhesive is usually used to hold the fiber securely, and a strain relief is secured to the rear. Once the adhesive sets, the fiber's end is polished to a mirror finish. Various polish profiles are used, depending on the type of fiber and the application. For single-mode fiber, fiber ends are typically polished with a slight curvature that makes the mated connectors touch only at their cores. This is called a *physical contact* (PC) polish. The curved surface may be polished at an angle, to make an *angled physical contact (APC)* connection. Such connections have higher loss than PC connections, but greatly reduced back reflection, because light that reflects from the angled surface leaks out of the fiber core. The resulting signal strength loss is called *gap loss*. APC fiber ends have low back reflection even when disconnected.

In the 1990s, terminating fiber optic cables was labor-intensive. The number of parts per connector, polishing of the fibers, and the need to oven-bake the epoxy in each connector made terminating fiber optic cables difficult. Today, many connectors types are on the market that offer easier, less labor-intensive ways of terminating cables. Some of the most popular connectors are pre-polished at the factory, and include a gel inside the connector. Those two steps help save money on labor, especially on large projects. A cleave is made at a required length, to get as close to the polished piece already inside the connector. The gel surrounds the point where the two pieces meet inside the connector for very little light loss.

Free-space Coupling

It is often necessary to align an optical fiber with another optical fiber, or with an optoelectronic device such as a light-emitting diode, a laser diode, or a modulator. This can involve either carefully aligning the fiber and placing it in contact with the device, or can use a lens to allow coupling over an air gap. Typically the size of the fiber mode is much larger than the size of the mode in a laser diode or a silicon optical chip. In this case a tapered or lensed fiber is used to match the fiber mode field distribution to that of the other element. The lens on the end of the fiber can be formed using polishing, laser cutting or fusion splicing.

In a laboratory environment, a bare fiber end is coupled using a fiber launch system, which uses a microscope objective lens to focus the light down to a fine point. A precision translation stage (micro-positioning table) is used to move the lens, fiber, or device to allow the coupling efficiency to be optimized. Fibers with a connector on the end make this process much simpler: the connector is simply plugged into a pre-aligned fiberoptic collimator, which contains a lens that is either accurately positioned with respect to the fiber, or is adjustable. To achieve the best injection efficiency into single-mode fiber, the direction, position, size and divergence of the beam must all be optimized. With good beams, 70 to 90% coupling efficiency can be achieved.

With properly polished single-mode fibers, the emitted beam has an almost perfect Gaussian shape—even in the far field—if a good lens is used. The lens needs to be large enough to support the full numerical aperture of the fiber, and must not introduce aberrations in the beam. Aspheric lenses are typically used.

Fiber Fuse

At high optical intensities, above 2 megawatts per square centimeter, when a fiber is subjected to a shock or is otherwise suddenly damaged, a *fiber fuse* can occur. The reflection from the damage vaporizes the fiber immediately before the break, and this new defect remains reflective so that the damage propagates back toward the transmitter at 1–3 meters per second (4–11 km/h, 2–8 mph). The open fiber control system, which ensures laser eye safety in the event of a broken fiber, can also effectively halt propagation of the fiber fuse. In situations, such as undersea cables, where high power levels might be used without the need for open fiber control, a "fiber fuse" protection device at the transmitter can break the circuit to keep damage to a minimum.

Chromatic Dispersion

The refractive index of fibers varies slightly with the frequency of light, and light sources are not perfectly monochromatic. Modulation of the light source to transmit a signal also slightly widens the frequency band of the transmitted light. This has the effect that, over long distances and at high modulation speeds, the different frequencies of light can take different times to arrive at the receiver, ultimately making the signal impossible to discern, and requiring extra repeaters. This problem can be overcome in a number of ways, including the use of a relatively short length of fiber that has the opposite refractive index gradient.

Fiber-optic Communication

Fiber-optic communication is a method of transmitting information from one place to another by sending pulses of light through an optical fiber. The light forms an electromagnetic carrier wave that is modulated to carry information. First developed in the 1970s, fiber-optics have revolutionized the telecommunications industry and have played a major role in the advent of the Information Age. Because of its advantages over electrical transmission, optical fibers have largely replaced copper wire communications in core networks in the developed world. Optical fiber is used by many telecommunications companies to transmit telephone signals, Internet communication, and cable television signals. Researchers at Bell Labs have reached internet speeds of over 100 petabit×kilometer per second using fiber-optic communication.

The process of communicating using fiber-optics involves the following basic steps:

1. creating the optical signal involving the use of a transmitter, usually from an electrical signal.
2. relaying the signal along the fiber, ensuring that the signal does not become too distorted or weak.
3. receiving the optical signal.
4. converting it into an electrical signal.

Applications

Optical fiber is used by many telecommunications companies to transmit telephone signals, Internet communication, and cable television signals. Due to much lower attenuation and interference, optical fiber has large advantages over existing copper wire in long-distance and high-demand applications. However, infrastructure development within cities was relatively difficult and time-consuming, and fiber-optic systems were complex and expensive to install and operate. Due to these difficulties, fiber-optic communication systems have primarily been installed in long-distance applications, where they can be used to their full transmission capacity, offsetting the increased cost. Since 2000, the prices for fiber-optic communications have dropped considerably.

The price for rolling out fiber to the home has currently become more cost-effective than that of rolling out a copper based network. Prices have dropped to $850 per subscriber in the US and lower in countries like The Netherlands, where digging costs are low and housing density is high.

Since 1990, when optical-amplification systems became commercially available, the telecommunications industry has laid a vast network of intercity and transoceanic fiber communication lines. By 2002, an intercontinental network of 250,000 km of submarine communications cable with a capacity of 2.56 Tb/s was completed, and although specific network capacities are privileged information, telecommunications investment reports indicate that network capacity has increased dramatically since 2004.

History

In 1880 Alexander Graham Bell and his assistant Charles Sumner Tainter created a very early precursor to fiber-optic communications, the Photophone, at Bell's newly established Volta Laboratory in Washington, D.C. Bell considered it his most important invention. The device allowed for the transmission of sound on a beam of light. On June 3, 1880, Bell conducted the world's first wireless telephone transmission between two buildings, some 213 meters apart. Due to its use of an atmospheric transmission medium, the Photophone would not prove practical until advances in laser and optical fiber technologies permitted the secure transport of light. The Photophone's first practical use came in military communication systems many decades later.

In 1954 Harold Hopkins and Narinder Singh Kapany showed that rolled fiber glass allowed light to be transmitted. Initially it was considered that the light can traverse in only straight medium.

In 1966 Charles K. Kao and George Hockham proposed optical fibers at STC Laboratories (STL) at Harlow, England, when they showed that the losses of 1,000 dB/km in existing glass (compared to 5-10 dB/km in coaxial cable) were due to contaminants, which could potentially be removed.

Optical fiber was successfully developed in 1970 by Corning Glass Works, with attenuation low enough for communication purposes (about 20 dB/km), and at the same time GaAs semiconductor lasers were developed that were compact and therefore suitable for transmitting light through fiber optic cables for long distances.

After a period of research starting from 1975, the first commercial fiber-optic communications system was developed, which operated at a wavelength around 0.8 μm and used GaAs semiconductor lasers. This first-generation system operated at a bit rate of 45 Mbps with repeater spacing of up to 10 km. Soon on 22 April 1977, General Telephone and Electronics sent the first live telephone traffic through fiber optics at a 6 Mbit/s throughput in Long Beach, California.

In October 1973, Corning Glass signed a development contract with CSELT and Pirelli aimed to test fiber optics in an urban environment: in September 1977, the second cable in this test series, named COS-2, was experimentally deployed in two lines (9 km) in Turin, for the first time in a big city, at a speed of 140 Mbit/s.

The second generation of fiber-optic communication was developed for commercial use in the early 1980s, operated at 1.3 μm, and used InGaAsP semiconductor lasers. These early systems were initially limited by multi mode fiber dispersion, and in 1981 the single-mode fiber was revealed to greatly improve system performance, however practical connectors capable of working with single mode fiber proved difficult to develop. In 1984, they had already developed a fiber optic cable that would help further their progress toward making fiber optic cables that would circle the globe. Canadian service provider SaskTel had completed construction of what was then the world's longest commercial fiberoptic network, which covered 3,268 km and linked 52 communities. By 1987, these systems were operating at bit rates of up to 1.7 Gb/s with repeater spacing up to 50 km.

The first transatlantic telephone cable to use optical fiber was TAT-8, based on Desurvire optimised laser amplification technology. It went into operation in 1988.

Third-generation fiber-optic systems operated at 1.55 μm and had losses of about 0.2 dB/km. This development was spurred by the discovery of Indium gallium arsenide and the development of the Indium Gallium Arsenide photodiode by Pearsall. Engineers overcame earlier difficulties with pulse-spreading at that wavelength using conventional InGaAsP semiconductor lasers. Scientists overcame this difficulty by using dispersion-shifted fibers designed to have minimal dispersion at 1.55 μm or by limiting the laser spectrum to a single longitudinal mode. These developments eventually allowed third-generation systems to operate commercially at 2.5 Gbit/s with repeater spacing in excess of 100 km.

The fourth generation of fiber-optic communication systems used optical amplification to reduce the need for repeaters and wavelength-division multiplexing to increase data capacity. These two improvements caused a revolution that resulted in the doubling of system capacity every six months starting in 1992 until a bit rate of 10 Tb/s was reached by 2001. In 2006 a bit-rate of 14 Tbit/s was reached over a single 160 km line using optical amplifiers.

The focus of development for the fifth generation of fiber-optic communications is on extending the wavelength range over which a WDM system can operate. The conventional wavelength window, known as the C band, covers the wavelength range 1.53-1.57 μm, and *dry fiber* has a low-loss window promising an extension of that range to 1.30-1.65 μm. Other developments include the concept of "optical solitons", pulses that preserve their shape by counteracting the effects of dispersion with the nonlinear effects of the fiber by using pulses of a specific shape.

In the late 1990s through 2000, industry promoters, and research companies such as KMI, and RHK predicted massive increases in demand for communications bandwidth due to increased use of the Internet, and commercialization of various bandwidth-intensive consumer services, such as video on demand. Internet protocol data traffic was increasing exponentially, at a faster rate than integrated circuit complexity had increased under Moore's Law. From the bust of the dot-com bubble through 2006, however, the main trend in the industry has been consolidation of firms and offshoring of manufacturing to reduce costs. Companies such as Verizon and AT&T have taken advantage of fiber-optic communications to deliver a variety of high-throughput data and broadband services to consumers' homes.

Technology

Modern fiber-optic communication systems generally include an optical transmitter to convert an electrical signal into an optical signal to send into the optical fiber, a cable containing bundles of multiple optical fibers that is routed through underground conduits and buildings, multiple kinds of amplifiers, and an optical receiver to recover the signal as an electrical signal. The information transmitted is typically digital information generated by computers, telephone systems, and cable television companies.

Transmitters

A GBIC module (shown here with its cover removed), is an optical and electrical transceiver. The electrical connector is at top right, and the optical connectors are at bottom left.

The most commonly used optical transmitters are semiconductor devices such as light-emitting diodes (LEDs) and laser diodes. The difference between LEDs and laser diodes is that LEDs produce incoherent light, while laser diodes produce coherent light. For use in optical communications, semiconductor optical transmitters must be designed to be compact, efficient, and reliable, while operating in an optimal wavelength range, and directly modulated at high frequencies.

In its simplest form, an LED is a forward-biased p-n junction, emitting light through spontaneous emission, a phenomenon referred to as electroluminescence. The emitted light is incoherent with a relatively wide spectral width of 30-60 nm. LED light transmission is also inefficient, with only about 1% of input power, or about 100 microwatts, eventually converted into launched power which has been coupled into the optical fiber. However, due to their relatively simple design, LEDs are very useful for low-cost applications.

Communications LEDs are most commonly made from Indium gallium arsenide phosphide (InGaAsP) or gallium arsenide (GaAs). Because InGaAsP LEDs operate at a longer wavelength than GaAs LEDs (1.3 micrometers vs. 0.81-0.87 micrometers), their output spectrum, while equivalent in energy is wider in wavelength terms by a factor of about 1.7. The large spectrum width of LEDs is subject to higher fiber dispersion, considerably limiting their bit rate-distance product (a common measure of usefulness). LEDs are suitable primarily for local-area-network applications with bit rates of 10-100 Mbit/s and transmission distances of a few kilometers. LEDs have also been developed that use several quantum wells to emit light at different wavelengths over a broad spectrum, and are currently in use for local-area WDM (Wavelength-Division Multiplexing) networks.

Today, LEDs have been largely superseded by VCSEL (Vertical Cavity Surface Emitting Laser) devices, which offer improved speed, power and spectral properties, at a similar cost. Common VCSEL devices couple well to multi mode fiber.

A semiconductor laser emits light through stimulated emission rather than spontaneous emission, which results in high output power (~100 mW) as well as other benefits related to the nature of coherent light. The output of a laser is relatively directional, allowing high coupling efficiency (~50 %) into single-mode fiber. The narrow spectral width also allows for high bit rates since it reduces the effect of chromatic dispersion. Furthermore, semiconductor lasers can be modulated directly at high frequencies because of short recombination time.

Commonly used classes of semiconductor laser transmitters used in fiber optics include VCSEL (Vertical-Cavity Surface-Emitting Laser), Fabry–Pérot and DFB (Distributed Feed Back).

Laser diodes are often directly modulated, that is the light output is controlled by a current applied directly to the device. For very high data rates or very long distance *links*, a laser source may be operated continuous wave, and the light modulated by an external device such as an electro-absorption modulator or Mach–Zehnder interferometer. External modulation increases the achievable link distance by eliminating laser chirp, which broadens the linewidth of directly modulated lasers, increasing the chromatic dispersion in the fiber.

A transceiver is a device combining a transmitter and a receiver in a single housing.

Receivers

The main component of an optical receiver is a photodetector, which converts light into electricity using the photoelectric effect. The primary photodetectors for telecommunications are made from Indium gallium arsenide The photodetector is typically a semiconductor-based photodiode. Several types of photodiodes include p-n photodiodes, p-i-n photodiodes, and avalanche photodiodes. Metal-semiconductor-metal (MSM) photodetectors are also used due to their suitability for circuit integration in regenerators and wavelength-division multiplexers.

Optical-electrical converters are typically coupled with a transimpedance amplifier and a limiting amplifier to produce a digital signal in the electrical domain from the incoming optical signal, which may be attenuated and distorted while passing through the channel. Further signal processing such as clock recovery from data (CDR) performed by a phase-locked loop may also be applied before the data is passed on.

Fiber Cable Types

A cable reel trailer with conduit that can carry optical fiber.

Multi-mode optical fiber in an underground service pit.

An optical fiber cable consists of a core, cladding, and a buffer (a protective outer coating), in which the cladding guides the light along the core by using the method of total internal reflection. The core and the cladding (which has a lower-refractive-index) are usually made of high-quality silica glass, although they can both be made of plastic as well. Connecting two optical fibers is done by fusion splicing or mechanical splicing and requires special skills and interconnection technology due to the microscopic precision required to align the fiber cores.

Two main types of optical fiber used in optic communications include multi-mode optical fibers and single-mode optical fibers. A multi-mode optical fiber has a larger core (≥ 50 micrometers), allowing less precise, cheaper transmitters and receivers to connect to it as well as cheaper connectors. However, a multi-mode fiber introduces multimode distortion, which often limits the bandwidth and length of the link. Furthermore, because of its higher dopant content, multi-mode fibers are usually expensive and exhibit higher attenuation. The core of a single-mode fiber is smaller (<10 micrometers) and requires more expensive components and interconnection methods, but allows much longer, higher-performance links.

In order to package fiber into a commercially viable product, it typically is protectively coated by using ultraviolet (UV), light-cured acrylate polymers, then terminated with optical fiber connectors, and finally assembled into a cable. After that, it can be laid in the ground and then run through the walls of a building and deployed aerially in a manner similar to copper cables. These fibers require less maintenance than common twisted pair wires, once they are deployed.

Specialized cables are used for long distance subsea data transmission, e.g. transatlantic communications cable. New (2011–2013) cables operated by commercial enterprises (Emerald Atlantis, Hibernia Atlantic) typically have four strands of fiber and cross the Atlantic (NYC-London) in 60-70ms. Cost of each such cable was about $300M in 2011. *source: The Chronicle Herald.*

Another common practice is to bundle many fiber optic strands within long-distance power transmission cable. This exploits power transmission rights of way effectively, ensures a power company can own and control the fiber required to monitor its own devices and lines, is effectively immune to tampering, and simplifies the deployment of smart grid technology.

Amplifier

The transmission distance of a fiber-optic communication system has traditionally been limited by fiber attenuation and by fiber distortion. By using opto-electronic repeaters, these problems have been eliminated. These repeaters convert the signal into an electrical signal, and then use a transmitter to send the signal again at a higher intensity than was received, thus counteracting the loss incurred in the previous segment. Because of the high complexity with modern wavelength-division multiplexed signals (including the fact that they had to be installed about once every 20 km), the cost of these repeaters is very high.

An alternative approach is to use an optical amplifier, which amplifies the optical signal directly without having to convert the signal into the electrical domain. It is made by doping a length of fiber with the rare-earth mineral erbium, and *pumping* it with light from a laser with a shorter wavelength than the communications signal (typically 980 nm). Amplifiers have largely replaced repeaters in new installations.

Wavelength-division Multiplexing

Wavelength-division multiplexing (WDM) is the practice of multiplying the available capacity of optical fibers through use of parallel channels, each channel on a dedicated wavelength of light. This requires a wavelength division multiplexer in the transmitting equipment and a demultiplexer (essentially a spectrometer) in the receiving equipment. Arrayed waveguide gratings are commonly used for multiplexing and demultiplexing in WDM. Using WDM technology now commercially available, the bandwidth of a fiber can be divided into as many as 160 channels to support a combined bit rate in the range of 1.6 Tbit/s.

Parameters

Bandwidth–distance Product

Because the effect of dispersion increases with the length of the fiber, a fiber transmission system is often characterized by its *bandwidth–distance product*, usually expressed in units of MHz·km. This value is a product of bandwidth and distance because there is a trade off between the bandwidth of the signal and the distance it can be carried. For example, a common multi-mode fiber with bandwidth–distance product of 500 MHz·km could carry a 500 MHz signal for 1 km or a 1000 MHz signal for 0.5 km.

Engineers are always looking at current limitations in order to improve fiber-optic communication, and several of these restrictions are currently being researched.

Record Speeds

Each fiber can carry many independent channels, each using a different wavelength of light (wavelength-division multiplexing). The net data rate (data rate without overhead bytes) per fiber is the per-channel data rate reduced by the FEC overhead, multiplied by the number of channels (usually up to eighty in commercial dense WDM systems as of 2008).

Year	Organization	Effective speed	WDM channels	Per channel speed	Distance
2009	Alcatel-Lucent	15 Tbit/s	155	100 Gbit/s	90 km
2010	NTT	69.1 Tbit/s	432	171 Gbit/s	240 km
2011	KIT	26 Tbit/s	1	26 Tbit/s	50 km
2011	NEC	101 Tbit/s	370	273 Gbit/s	165 km
2012	NEC, Corning	1.05 Petabit/s	12 core fiber		52.4 km

While the physical limitations of electrical cable prevent speeds in excess of 10 Gigabits per second, the physical limitations of fiber optics have not yet been reached.In 2013, *New Scientist* reported that a team at the University of Southampton had achieved a throughput of 73.7 Tbit per second, with the signal traveling at 99.7% the speed of light through a hollow-core photonic crystal fiber.

Dispersion

For modern glass optical fiber, the maximum transmission distance is limited not by direct material absorption but by several types of dispersion, or spreading of optical pulses as they travel along the fiber. Dispersion in optical fibers is caused by a variety of factors. Intermodal dispersion, caused by the different axial speeds of different transverse modes, limits the performance of multi-mode fiber. Because single-mode fiber supports only one transverse mode, intermodal dispersion is eliminated.

In single-mode fiber performance is primarily limited by chromatic dispersion (also called group velocity dispersion), which occurs because the index of the glass varies slightly depending on the wavelength of the light, and light from real optical transmitters necessarily has nonzero spectral width (due to modulation). Polarization mode dispersion, another source of limitation, occurs because although the single-mode fiber can sustain only one transverse mode, it can carry this mode with two different polarizations, and slight imperfections or distortions in a fiber can alter the propagation velocities for the two polarizations. This phenomenon is called fiber birefringence and can be counteracted by polarization-maintaining optical fiber. Dispersion limits the bandwidth of the fiber because the spreading optical pulse limits the rate that pulses can follow one another on the fiber and still be distinguishable at the receiver.

Some dispersion, notably chromatic dispersion, can be removed by a 'dispersion compensator'. This works by using a specially prepared length of fiber that has the opposite dispersion to that induced by the transmission fiber, and this sharpens the pulse so that it can be correctly decoded by the electronics.

Attenuation

Fiber attenuation, which necessitates the use of amplification systems, is caused by a combination of material absorption, Rayleigh scattering, Mie scattering, and connection losses. Although material absorption for pure silica is only around 0.03 dB/km (modern fiber has attenuation around 0.3 dB/km), impurities in the original optical fibers caused attenuation of about 1000 dB/km. Other forms of attenuation are caused by physical stresses to the fiber, microscopic fluctuations in density, and imperfect splicing techniques.

Transmission Windows

Each effect that contributes to attenuation and dispersion depends on the optical wavelength. There are wavelength bands (or windows) where these effects are weakest, and these are the most favorable for transmission. These windows have been standardized, and the currently defined bands are the following:

Band	Description	Wavelength Range
O band	original	1260 to 1360 nm
E band	extended	1360 to 1460 nm
S band	short wavelengths	1460 to 1530 nm
C band	conventional ("erbium window")	1530 to 1565 nm
L band	long wavelengths	1565 to 1625 nm
U band	ultralong wavelengths	1625 to 1675 nm

Note that this table shows that current technology has managed to bridge the second and third windows that were originally disjoint.

Historically, there was a window used below the O band, called the first window, at 800-900 nm; however, losses are high in this region so this window is used primarily for short-distance communications. The current lower windows (O and E) around 1300 nm have much lower losses. This region has zero dispersion. The middle windows (S and C) around 1500 nm are the most widely used. This region has the lowest attenuation losses and achieves the longest range. It does have some dispersion, so dispersion compensator devices are used to remove this.

Regeneration

When a communications link must span a larger distance than existing fiber-optic technology is capable of, the signal must be *regenerated* at intermediate points in the link by optical communications repeaters. Repeaters add substantial cost to a communication system, and so system designers attempt to minimize their use.

Recent advances in fiber and optical communications technology have reduced signal degradation so far that *regeneration* of the optical signal is only needed over distances of hundreds of kilometers. This has greatly reduced the cost of optical networking, particularly over undersea spans where the cost and reliability of repeaters is one of the key factors determining the performance of the whole cable system. The main advances contributing to these performance improvements are dispersion management, which seeks to balance the effects of dispersion against non-linearity; and solitons, which use nonlinear effects in the fiber to enable dispersion-free propagation over long distances.

Last Mile

Although fiber-optic systems excel in high-bandwidth applications, optical fiber has been slow to achieve its goal of fiber to the premises or to solve the last mile problem. However, as bandwidth demand increases, more and more progress towards this goal can be observed. In Japan, for instance EPON has largely replaced DSL as a broadband Internet source. South Korea's KT also provides a service called FTTH (Fiber To The Home), which provides fiber-optic connections to the subscriber's home. The largest FTTH deployments are in Japan, South Korea, and China. Singapore started implementation of their all-fiber Next Generation Nationwide Broadband Network (Next Gen NBN), which is slated for completion in 2012 and is being installed by OpenNet. Since they began rolling out services in September 2010, network coverage in Singapore has reached 85% nationwide.

In the US, Verizon Communications provides a FTTH service called FiOS to select high-ARPU (Average Revenue Per User) markets within its existing territory. The other major surviving ILEC (or Incumbent Local Exchange Carrier), AT&T, uses a FTTN (Fiber To The Node) service called U-verse with twisted-pair to the home. Their MSO competitors employ FTTN with coax using HFC. All of the major access networks use fiber for the bulk of the distance from the service provider's network to the customer.

Also in the US, Wilson Utilities, located in Wilson, North Carolina, has implemented FTTH and has successfully achieved 1 gigabit fiber to the home. This was implemented in late 2013. Wilson Utilities first rolled out their FTTH (Fiber to the Home) in 2012 with speeds offerings of 20/40/60/100 megabits per second. Their service is referred to as GreenLight.

Some other small cities in the US, such as Morristown, TN, have had their local utility company, Morristown Utility Systems in this case, deploy FTTH, offering symmetric gigabit speeds to each subscriber (though most are 50/50 or 100/100 mbit). It's called MUS Fibernet. AT&T and others have aggressively sought legislation at the state level to prevent further competition from municipalities, despite their low investment in rural areas.

The globally dominant access network technology is EPON (Ethernet Passive Optical Network). In Europe, and among telcos in the United States, BPON (ATM-based Broadband PON) and GPON (Gigabit PON) had roots in the FSAN (Full Service Access Network) and ITU-T standards organizations under their control.

Comparison with Electrical Transmission

A mobile fiber optic splice lab used to access and splice underground cables.

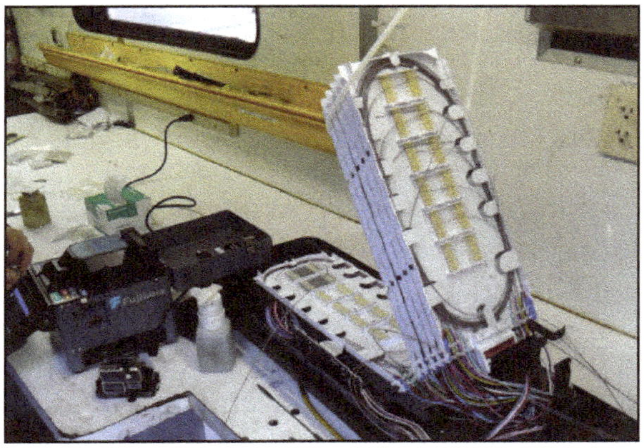

An underground fiber optic splice enclosure opened up.

The choice between optical fiber and electrical (or copper) transmission for a particular system is made based on a number of trade-offs. Optical fiber is generally chosen for systems requiring higher bandwidth or spanning longer distances than electrical cabling can accommodate.

The main benefits of fiber are its exceptionally low loss (allowing long distances between amplifiers/repeaters), its absence of ground currents and other parasite signal and power issues common to long parallel electric conductor runs (due to its reliance on light rather than electricity for transmission, and the dielectric nature of fiber optic), and its inherently high data-carrying capacity. Thousands of electrical links would be required to replace a single high bandwidth fiber cable. Another benefit of fibers is that even when run alongside each other for long distances, fiber cables experience effectively no crosstalk, in contrast to some types of electrical transmission lines. Fiber can be installed in areas with high electromagnetic interference (EMI), such as alongside utility lines, power lines, and railroad tracks. Nonmetallic all-dielectric cables are also ideal for areas of high lightning-strike incidence.

For comparison, while single-line, voice-grade copper systems longer than a couple of kilometers require in-line signal repeaters for satisfactory performance; it is not unusual for optical systems to go over 100 kilometers (62 mi), with no active or passive processing. Single-mode fiber cables are commonly available in 12 km lengths, minimizing the number of splices required over a long cable run. Multi-mode fiber is available in lengths up to 4 km, although industrial standards only mandate 2 km unbroken runs.

In short distance and relatively low bandwidth applications, electrical transmission is often preferred because of its

- Lower material cost, where large quantities are not required.
- Lower cost of transmitters and receivers.
- Capability to carry electrical power as well as signals (in appropriately designed cables).
- Ease of operating transducers in linear mode.
- Crosstalk from nearby cables and other parasitical unwanted signals increase profits from replacement and mitigation devices.

Optical fibers are more difficult and expensive to splice than electrical conductors. And at higher powers, optical fibers are susceptible to fiber fuse, resulting in catastrophic destruction of the fiber core and damage to transmission components.

Because of these benefits of electrical transmission, optical communication is not common in short box-to-box, backplane, or chip-to-chip applications; however, optical systems on those scales have been demonstrated in the laboratory.

In certain situations fiber may be used even for short distance or low bandwidth applications, due to other important features:

- Immunity to electromagnetic interference, including nuclear electromagnetic pulses.
- High electrical resistance, making it safe to use near high-voltage equipment or between areas with different earth potentials.
- Lighter weight—important, for example, in aircraft.
- No sparks—important in flammable or explosive gas environments.
- Not electromagnetically radiating, and difficult to tap without disrupting the signal—important in high-security environments.
- Much smaller cable size—important where pathway is limited, such as networking an existing building, where smaller channels can be drilled and space can be saved in existing cable ducts and trays.
- Resistance to corrosion due to non-metallic transmission medium

Optical fiber cables can be installed in buildings with the same equipment that is used to install copper and coaxial cables, with some modifications due to the small size and limited pull tension and bend radius of optical cables. Optical cables can typically be installed in duct systems in spans of 6000 meters or more depending on the duct's condition, layout of the duct system, and installation technique. Longer cables can be coiled at an intermediate point and pulled farther into the duct system as necessary.

Governing Standards

In order for various manufacturers to be able to develop components that function compatibly in fiber optic communication systems, a number of standards have been developed. The International Telecommunications Union publishes several standards related to the characteristics and performance of fibers themselves, including

- ITU-T G.651, "Characteristics of a 50/125 μm multimode graded index optical fibre cable".
- ITU-T G.652, "Characteristics of a single-mode optical fibre cable".

Other standards specify performance criteria for fiber, transmitters, and receivers to be used together in conforming systems. Some of these standards are:

- 100 Gigabit Ethernet.
- 10 Gigabit Ethernet.
- Fibre Channel.
- Gigabit Ethernet.
- HIPPI.
- Synchronous Digital Hierarchy.
- Synchronous Optical Networking.
- Optical Transport Network (OTN).

TOSLINK is the most common format for digital audio cable using plastic optical fiber to connect digital sources to digital receivers.

Fiber Optic Sensor

A fiber optic sensor is a sensor that uses optical fiber either as the sensing element ("intrinsic sensors"), or as a means of relaying signals from a remote sensor to the electronics that process the signals ("extrinsic sensors"). Fibers have many uses in remote sensing. Depending on the application, fiber may be used because of its small size, or because no electrical power is needed at the remote location, or because many sensors can be multiplexed along the length of a fiber by using light wavelength shift for each sensor, or by sensing the time delay as light passes along the fiber through each sensor. Time delay can be determined using a device such as an optical time-domain reflectometer and wavelength shift can be calculated using an instrument implementing optical frequency domain reflectometry.

Fiber optic sensors are also immune to electromagnetic interference, and do not conduct electricity so they can be used in places where there is high voltage electricity or flammable material such as jet fuel. Fiber optic sensors can be designed to withstand high temperatures as well.

Intrinsic Sensors

Optical fibers can be used as sensors to measure strain,temperature, pressure and other quantities by modifying a fiber so that the quantity to be measured modulates the intensity, phase, polarization, wavelength or transit time of light in the fiber. Sensors that vary the intensity of light are the simplest, since only a simple source and detector are required. A particularly useful feature of intrinsic fiber optic sensors is that they can, if required, provide distributed sensing over very large distances.

Temperature can be measured by using a fiber that has evanescent loss that varies with temperature, or by analyzing the Raman scattering of the optical fiber. Electrical voltage can be sensed by nonlinear optical effects in specially-doped fiber, which alter the polarization of light as a function of voltage or electric field. Angle measurement sensors can be based on the Sagnac effect.

Special fibers like long-period fiber grating (LPG) optical fibers can be used for direction recognition . Photonics Research Group of Aston University in UK has some publications on vectorial bend sensor applications.

Optical fibers are used as hydrophones for seismic and sonar applications. Hydrophone systems with more than one hundred sensors per fiber cable have been developed. Hydrophone sensor systems are used by the oil industry as well as a few countries' navies. Both bottom-mounted hydrophone arrays and towed streamer systems are in use. The German company Sennheiser developed a laser microphone for use with optical fibers.

A fiber optic microphone and fiber-optic based headphone are useful in areas with strong electrical or magnetic fields, such as communication amongst the team of people working on a patient inside a magnetic resonance imaging (MRI) machine during MRI-guided surgery.

Optical fiber sensors for temperature and pressure have been developed for downhole measurement in oil wells. The fiber optic sensor is well suited for this environment as it functions at temperatures too high for semiconductor sensors (distributed temperature sensing).

Optical fibers can be made into interferometric sensors such as fiber optic gyroscopes, which are used in the Boeing 767 and in some car models (for navigation purposes). They are also used to make hydrogen sensors.

Fiber-optic sensors have been developed to measure co-located temperature and strain simultaneously with very high accuracy using fiber Bragg gratings. This is particularly useful when acquiring information from small or complex structures. Fiber Bragg grating sensors are also particularly well suited for remote monitoring, and they can be interrogated 250 km away from the monitoring station using an optical fiber cable. Brillouin scattering effects can also be used to detect strain and temperature over large distances (20–120 kilometers).

Other Examples

A fiber-optic AC/DC voltage sensor in the middle and high voltage range (100–2000 V) can be created by inducing measurable amounts of Kerr nonlinearity in single mode optical fiber by exposing a calculated length of fiber to the external electric field. The measurement technique is based on polarimetric detection and high accuracy is achieved in a hostile industrial environment.

High frequency (5 MHz–1 GHz) electromagnetic fields can be detected by induced nonlinear effects in fiber with a suitable structure. The fiber used is designed such that the Faraday and Kerr effects cause considerable phase change in the presence of the external field. With appropriate sensor design, this type of fiber can be used to measure different electrical and magnetic quantities and different internal parameters of fiber material.

Electrical power can be measured in a fiber by using a structured bulk fiber ampere sensor coupled with proper signal processing in a polarimetric detection scheme. Experiments have been carried out in support of the technique.

Fiber-optic sensors are used in electrical switchgear to transmit light from an electrical arc flash to a digital protective relay to enable fast tripping of a breaker to reduce the energy in the arc blast.

Fiber Bragg grating based fiber optic sensors significantly enhance performance, efficiency and safety in several industries. With FBG integrated technology, sensors can provide detailed analysis and comprehensive reports on insights with very high resolution. These type of sensors are used extensively in several industries like telecommunication, automotive, aerospace, energy, etc.Fiber Bragg gratings are sensitive to the static pressure, mechanical tension and compression and fiber temperature changes.

The efficiency of fiber Bragg grating based fiber optic sensors can be provided by means of central wavelength adjustment of light emitting source in accordance with the current Bragg gratings reflection spectra.

Extrinsic Sensors

Extrinsic fiber optic sensors use an optical fiber cable, normally a multimode one, to transmit modulated light from either a non-fiber optical sensor, or an electronic sensor connected to an optical transmitter. A major benefit of extrinsic sensors is their ability to reach places which are otherwise inaccessible. An example is the measurement of temperature inside aircraft jet engines by using a fiber to transmit radiation into a radiation pyrometer located outside the engine. Extrinsic sensors can also be used in the same way to measure the internal temperature of electrical transformers, where the extreme electromagnetic fields present make other measurement techniques impossible.

Extrinsic fiber optic sensors provide excellent protection of measurement signals against noise corruption. Unfortunately, many conventional sensors produce electrical output which must be converted into an optical signal for use with fiber. For example, in the case of a platinum resistance thermometer, the temperature changes are translated into resistance changes. The PRT must therefore have an electrical power supply. The modulated voltage level at the output of the PRT can then be injected into the optical fiber via the usual type of transmitter. This complicates the measurement process and means that low-voltage power cables must be routed to the transducer.

Extrinsic sensors are used to measure vibration, rotation, displacement, velocity, acceleration, torque, and temperature.

Free-space Optical Communication

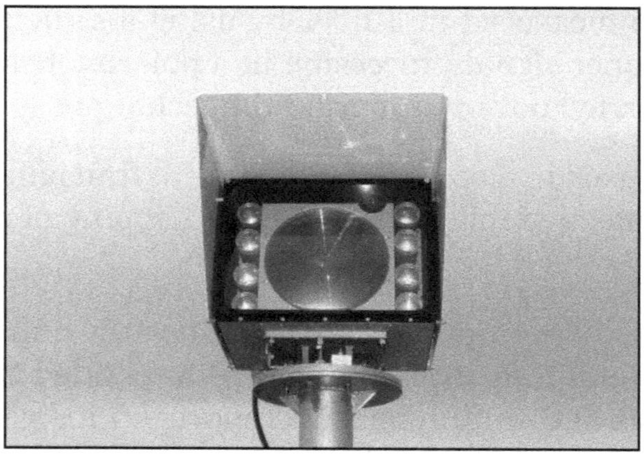

An 8-beam free space optics laser link, rated for 1 Gbit/s. The receptor is the large disc in the middle, the transmitters the smaller ones. At the top right corner is a monocular forassisting the alignment of the two heads.

Free-space optical communication (FSO) is an optical communication technology that uses light propagating in free space to wirelessly transmit data for telecommunications or computer networking. "Free space" means air, outer space, vacuum, or something similar. This contrasts with using solids such as optical fiber cable or an optical transmission line. The technology is useful where the physical connections are impractical due to high costs or other considerations.

History

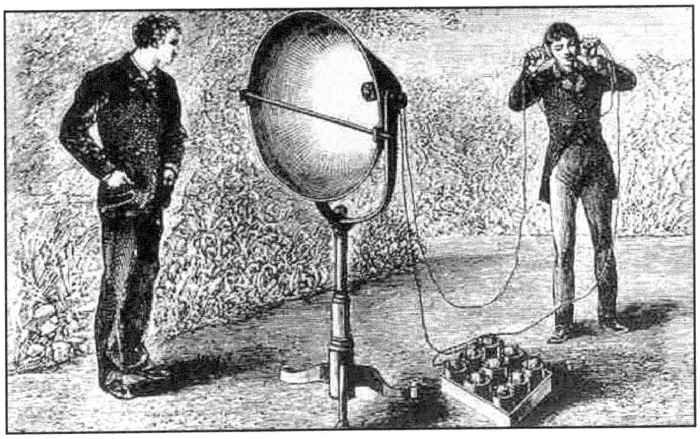

A photophone receiver and headset, one half of Bell and Tainter's optical telecommunication system of 1880.

Optical communications, in various forms, have been used for thousands of years. The Ancient Greeks used a coded alphabetic system of signalling with torches developed by Cleoxenus, Democleitus and Polybius. In the modern era, semaphores and wireless solar telegraphs called heliographs were developed, using coded signals to communicate with their recipients.

In 1880 Alexander Graham Bell and his assistant Charles Sumner Tainter created the Photophone, at Bell's newly established Volta Laboratory in Washington, DC. Bell considered it his most important invention. The device allowed for the transmission of sound on a beam of light. On June 3, 1880, Bell conducted the world's first wireless telephone transmission between two buildings, some 213 meters (700 feet) apart.

Its first practical use came in military communication systems many decades later, first for optical telegraphy. German colonial troops used Heliograph telegraphy transmitters during the 1904/05 Herero Genocide in German South-West Africa (today's Namibia) as did British, French, US or Ottoman signals.

During the trench warfare of World War I when wire communications were often cut, German signals used three types of optical Morse transmitters called *Blinkgerät*, the intermediate type for distances of up to 4 km (2.5 miles) at daylight and of up to 8 km (5 miles) at night, using red filters for undetected communications. Optical telephone communications were tested at the end of the war, but not introduced at troop level. In addition, special blinkgeräts were used for communication with airplanes, balloons, and tanks, with varying success.

WW I German Blinkgerät.

A major technological step was to replace the Morse code by modulating optical waves in speech transmission. Carl Zeiss Jena developed the *Lichtsprechgerät 80/80* (literal translation: optical speaking device) that the German army used in their World War II anti-aircraft defense units, or in bunkers at the Atlantic Wall.

The invention of lasers in the 1960s revolutionized free space optics. Military organizations were particularly interested and boosted their development. However the technology lost market momentum when the installation of optical fiber networks for civilian uses was at its peak.

Many simple and inexpensive consumer remote controls use low-speed communication using infrared (IR) light. This is known as consumer IR technologies.

A recently declassified 1987 Pentagon report reveals free-space lasers have been mounted on Israeli F-15 fighter jets for the purposes of surveillance, missile-tracking, and targeted weaponry.

Usage and Technologies

Free-space point-to-point optical links can be implemented using infrared laser light, although low-data-rate communication over short distances is possible using LEDs. Infrared Data Association (IrDA) technology is a very simple form of free-space optical communications. On the communications side the FSO technology is considered as a part of the Optical Wireless Communications applications. Free-space optics can be used for communications between spacecraft.

Current Market Demands

The demand for a high-speed (10 GBps+) and long range (3 – 5 km) FSO system is apparent in the market place.

- In 2008, MRV Communications introduced a free-space optics (FSO)-based system with a data rate of 10GB/s initially claiming a distance of 2 km at high availability. This equipment is no longer available; before end-of-life, the product's useful distance was changed down to 350m.
- In 2013, the company MOSTCOM started to serially produce a new wireless communication system that also had a data rate of 10Gb/s as well as an improved range of up to 2.5 km, but to get to 99.99% up-time the designers used an RF hybrid solution, meaning the data rate drops to extremely low levels during atmospheric disturbances (typically down to 10MB/s).
- LightPointe offers many similar hybrid solutions to MOSTCOM's offering.

Useful Distances

The reliability of FSO units has always been a problem for commercial telecommunications. Consistently, studies find too many dropped packets and signal errors over small ranges (400 to 500 meters). This is from both independent studies, such as in the Czech republic, as well as formal internal nationwide studies, such as one conducted by MRV FSO staff. Military based studies consistently produce longer estimates for reliability, projecting the maximum range for terrestrial links is of the order of 2 to 3 km (1.2 to 1.9 mi). All studies agree the stability and quality of the link is highly dependent on atmospheric factors such as rain, fog, dust and heat.

Extending the Useful Distance

The main reason terrestrial communications have been limited to non-commercial telecommunications functions is fog. Fog consistently keeps FSO laser links over 500 meters from achieving a year-round bit error rate of 99.999%. Several entities are continually attempting to overcome these key disadvantages to FSO communications and field a system with a better quality of service. DARPA has sponsored over $130 million USD in research towards this effort, with the ORCA and ORCLE programs.

Other non-government groups are fielding tests to evaluate different technologies that some claim have the ability to address key FSO adoption challenges. As of October 2014, none have fielded a working system that addresses the most common atmospheric events.

DARPA ORCA Official Concept Art created circa 2008.

FSO research from 1998-2006 in the private sector totaled $407.1 million, divided primarily among 4 start-up companies. All four failed to deliver products that would meet telecommunications quality and distance standards:

- Terabeam received approximately $226 million in funding. AT&T and Lucent backed this attempt. The work ultimately failed, and the company reorganized in 2004.

- AirFiber received $96.1 million in funding, and never solved the weather issue. They sold out to MRV communications in 2003, and MRV sold their FSO units until 2012 when the end-of-life was abruptly announced for the Terescope series.

- LightPointe Communications received $76 million in start-up funds, and eventually reorganized to sell hybrid FSO-RF units to overcome the weather-based challenges.

- The Maxima Corporation published its operating theory in Science (magazine), and received $9 million in funding before permanently shutting down. No known spin-off or purchase followed this effort.

One private company published a paper on Nov 20,2014, claiming they had achieved commercial reliability (99.999% availability) in extreme fog. There is no indication this product is currently commercially available.

Extraterrestrial

The massive advantages of laser communication in space have multiple space agencies racing to develop a stable space communication platform, with many significant demonstrations and achievements. To date (18 December 2014), *no laser communication system is in use in space.*

Demonstrations in Space:

The first gigabit laser-based communication was achieved by the European Space Agency and called the European Data Relay System (EDRS) on November 28, 2014. The initial images have just been demonstrated, and a working system is expected to be in place in the 2015-2016 time frame.

NASA's OPALS announced a breakthrough in space-to-ground communication December 9, 2014, uploading 175 megabytes in 3.5 seconds. Their system is also able to re-acquire tracking after the signal was lost due to cloud cover.

In January 2013, NASA used lasers to beam an image of the Mona Lisa to the Lunar Reconnaissance Orbiter roughly 390,000 km (240,000 mi) away. To compensate for atmospheric interference, an error correction code algorithm similar to that used in CDs was implemented.

A two-way distance record for communication was set by the Mercury laser altimeter instrument aboard the MESSENGER spacecraft, and was able to communicate across a distance of 24 million km (15 million miles), as the craft neared Earth on a fly-by in May, 2005. The previous record had been set with a one-way detection of laser light from Earth, by the Galileo probe, of 6 million km in 1992. Quote from Laser Communication in Space Demonstrations (EDRS).

LEDs

RONJA is a free implementation of FSO using high-intensity LEDs.

In 2001, Twibright Labs released Ronja Metropolis, an open source DIY 10Mbit/s full duplex LED FSO over 1.4 km In 2004, a Visible Light Communication Consortium was formed in Japan. This was based on work from researchers that used a white LED-based space lighting system for indoor local area network (LAN) communications. These systems present advantages over traditional UHF RF-based systems from improved isolation between systems, the size and cost of receivers/transmitters, RF licensing laws and by combining space lighting and communication into the same system. In January 2009 a task force for visible light communication was formed by the

Institute of Electrical and Electronics Engineers working group for wireless personal area network standards known as IEEE 802.15.7. A trial was announced in 2010 in St. Cloud, Minnesota.

Amateur radio operators have achieved significantly farther distances using incoherent sources of light from high-intensity LEDs. One reported 173 miles (278 km) in 2007. However, physical limitations of the equipment used limited bandwidths to about 4 kHz. The high sensitivities required of the detector to cover such distances made the internal capacitance of the photodiode used a dominant factor in the high-impedance amplifier which followed it, thus naturally forming a low-pass filter with a cut-off frequency in the 4 kHz range. From the other side use of lasers radiation source allows to reach very high data rates which are comparable to fiber communications.

Projected data rates and future data rate claims vary. A low-cost white LED (GaN-phosphor) which could be used for space lighting can typically be modulated up to 20 MHz. Data rates of over 100 Mbit/s can be easily achieved using efficient modulation schemes and Siemens claimed to have achieved over 500 Mbit/s in 2010. Research published in 2009 used a similar system for traffic control of automated vehicles with LED traffic lights.

In September 2013, pureLiFi, the Edinburgh start-up working on Li-Fi, also demonstrated high speed point-to-point connectivity using any off-the-shelf LED light bulb. In previous work, high bandwidth specialist LEDs have been used to achieve the high data rates. The new system, the Li-1st, maximizes the available optical bandwidth for any LED device, thereby reducing the cost and improving the performance of deploying indoor FSO systems.

Engineering Details

Typically, best use scenarios for this technology are:

- LAN-to-LAN connections on campuses at Fast Ethernet or Gigabit Ethernet speeds.
- LAN-to-LAN connections in a city, a metropolitan area network.
- To cross a public road or other barriers which the sender and receiver do not own.
- Speedy service delivery of high-bandwidth access to optical fiber networks.
- Converged Voice-Data-Connection.
- Temporary network installation (for events or other purposes).

- Reestablish high-speed connection quickly (disaster recovery).
- As an alternative or upgrade add-on to existing wireless technologies.
    - Especially powerful in combination with auto aiming systems, this way you could power moving cars or you can power your laptop while you move or use auto-aiming nodes to create a network with other nodes.
- As a safety add-on for important fiber connections (redundancy).
- For communications between spacecraft, including elements of a satellite constellation.
- For inter- and intra -chip communication.

The light beam can be very narrow, which makes FSO hard to intercept, improving security. In any case, it is comparatively easy to encrypt any data traveling across the FSO connection for additional security. FSO provides vastly improved electromagnetic interference (EMI) behavior compared to using microwaves.

Technical Advantages

- Ease of deployment.
- Can be used to power devices.
- License-free long-range operation (in contrast with radio communication).
- High bit rates.
- Low bit error rates.
- Immunity to electromagnetic interference.
- Full duplex operation.
- Protocol transparency.
- Increased security when working with narrow beam(s).
- No Fresnel zone necessary.
- Reference open source implementation.

Range Limiting Factors

For terrestrial applications, the principal limiting factors are:

- Fog (10..~100 dB/km attenuation).
- Beam dispersion.

- Atmospheric absorption.
- Rain.
- Snow.
- Terrestrial scintillation.
- Interference from background light sources (including the Sun).
- Shadowing.
- Pointing stability in wind.
- Pollution / smog.

These factors cause an attenuated receiver signal and lead to higher bit error ratio (BER). To overcome these issues, vendors found some solutions, like multi-beam or multi-path architectures, which use more than one sender and more than one receiver. Some state-of-the-art devices also have larger fade margin (extra power, reserved for rain, smog, fog). To keep an eye-safe environment, good FSO systems have a limited laser power density and support laser classes 1 or 1M. Atmospheric and fog attenuation, which are exponential in nature, limit practical range of FSO devices to several kilometres.

Optical Wireless Communication

Optical wireless communications (OWC) is a form of optical communication in which unguided visible, infrared (IR), or ultraviolet (UV) light is used to carry a signal.

OWC systems operating in the visible band (390–750 nm) are commonly referred to as visible light communication (VLC). VLC systems take advantage of light emitting diodes (LEDs) which can be pulsed at very high speeds without noticeable effect on the lighting output and human eye. VLC can be possibly used in a wide range of applications including wireless local area networks, wireless personal area networks and vehicular networks among others. On the other hand, terrestrial point-to-point OWC systems, also known as the free space optical (FSO) systems, operate at the near IR frequencies (750–1600 nm). These systems typically use laser transmitters and offer a cost-effective protocol-transparent link with high data rates, i.e., 10 Gbit/s per wavelength, and provide a potential solution for the backhaul bottleneck. There has also been a growing interest on ultraviolet communication (UVC) as a result of recent progress in solid state optical sources/detectors operating within solar-blind UV spectrum (200–280 nm). In this so-called deep UV band, solar radiation is negligible at the ground level and this makes possible the design of photon-counting detectors with wide field-of-view receivers that increase the received energy with little additional background noise. Such designs are particularly useful for outdoor non-line-of-sight configurations to support low power short-range UVC such as in wireless sensor and ad-hoc networks.

History

The proliferation of wireless communications stands out as one of the most significant phenomena in the history of technology. Wireless technologies have become essential much more quickly during the last four decades and they will be a key element of society progress for the foreseeable future. The radio-frequency (RF) technologies wide-scale deployment has become the key factor to the wireless devices and systems expansion. However, the electromagnetic spectrum where the wireless systems are deployed is limited in capacity and costly according to its exclusive licenses of exploitation. With the raise of data heavy wireless communications, the demand for RF spectrum is outstripping supply and they become to consider other viable options for wireless communication using the upper parts of the electromagnetic spectrum not just RF.

Optical wireless communication (OWC) refers to transmission in unguided propagation media through the use of optical carriers, i.e., visible, infrared (IR), and ultraviolet (UV) band. Signalling through beacon fires, smoke, ship flags and semaphore telegraph can be considered the historical forms of OWC. Sunlight has been also used for long distance signalling since very early times. The earliest use of sunlight for communication purposes is attributed to ancient Greeks and Romans who used their polished shields to send signals by reflecting sunlight during battles. In 1810, Carl Friedrich Gauss invented the heliograph which involves a pair of mirrors to direct a controlled beam of sunlight to a distant station. Although the original heliograph was designed for geodetic survey, it was used extensively for military purposes during the late 19th and early 20th century. In 1880, Alexander Graham Bell invented the photophone, known as the world's first wireless telephone system.

The military interest on photophone however continued. For example, in 1935, the German Army developed a photophone where a tungsten filament lamp with an IR transmitting filter was used as a light source. Also, American and German military laboratories continued the development of high pressure arc lamps for optical communication until the 1950s. In modern sense, OWC uses either lasers or light emitting diodes (LEDs) as transmitters. In 1962, MIT Lincoln Labs built an experimental OWC link using a light emitting GaAs diode and was able to transmit TV signals over a distance of 30 miles. After the invention of laser, OWC was envisioned to be the main deployment area for lasers and many trials were conducted using different types of lasers and modulation schemes. However, the results were in general disappointing due to large divergence of laser beams and the inability to cope with atmospheric effects. With the development of low-loss fiber optics in the 1970s, they became the obvious choice for long distance optical transmission and shifted the focus away from OWC systems.

Current Status

Over the decades, the interest in OWC remained mainly limited to covert military applications, and space applications including inter-satellite and deep-space links. OWC's mass market penetration has been so far limited with the exception of IrDA which became a highly successful wireless short-range transmission solution. Development of novel and efficient wireless technologies for a range of transmission links is essential for building future heterogeneous communication networks to support a wide range of service types with various traffic patterns and to meet the ever-increasing demands for higher data rates. Variations of OWC can be potentially employed in a diverse range of communication applications ranging from optical interconnects within integrated circuits through outdoor inter-building links to satellite communications.

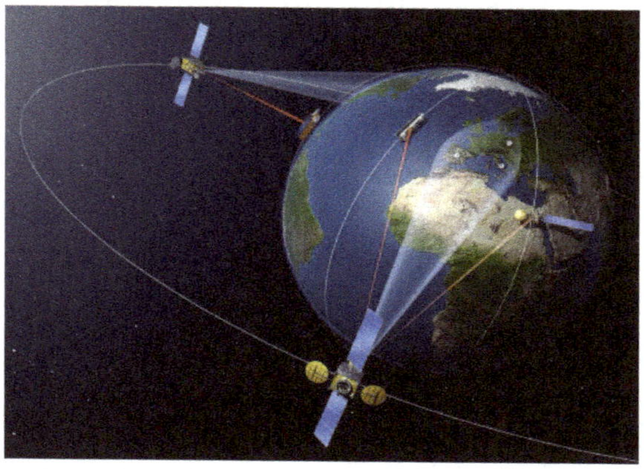

Photo: ESA; Example of possible ultra-long OWC on inter-satellite European Data Relay Satellite (EDRS) system.

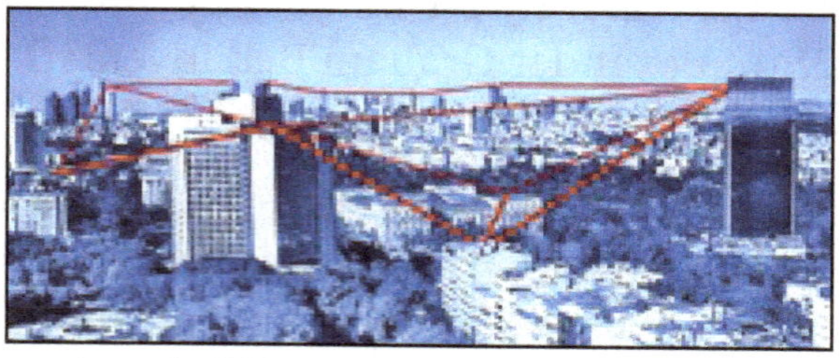

OWC long range inter-buildings communications idea for Istanbul Skyline.

Applications

Based on the transmission range, OWC can be studied in five categories:

- Ultra-short range OWC: chip-to-chip communications in stacked and closely packed multi-chip packages.
- Short range OWC: wireless body area network (WBAN) and wireless personal area network (WPAN) applications under standard IEEE 802.15.7, underwater communications.

- Medium range OWC: indoor IR and visible light communications (VLC) for wireless local area networks (WLANs) and inter-vehicular and vehicle-to-infrastructure communications.
- Long range OWC,: inter-building connections, also called Free-Space Optical Communications (FSO).
- Ultra-long range OWC: inter-satellite links.

Recent Trends

- In January 2015, IEEE 802.15 formed a Task Group to write a revision to IEEE 802.15.7-2011 that accommodates infrared and near ultraviolet wavelengths, in addition to visible light, and adds options such as Optical Camera Communications and LiFi.
- At long range OWC applications a 1 Gbit/s - 60 km range link between ground to aircraft at 800 km/h speed has been demonstrated, "Extreme Test for the ViaLight Laser Communication Terminal MLT-20 – Optical Downlink from a Jet Aircraft at 800 km/h", DLR and EADS December 2013.
- On consumer devices and short-range OWC applications on phones; Charge and receive data with light at your smartphone: TCL Communication/ALCATEL ONETOUCH and Sunpartner Technologies announces the first fully integrated solar smartphone. March 2014.
- On ultra-long range OWC applications the NASA's Lunar Laser Communication Demonstration (LLCD) transmitted data from lunar orbit to Earth at a rate of 622 Megabits-per-second (Mbps), November 2013.
- The Next Generation of OWC / Visible Light Communications demonstrated 10 Mb/s transmission with Polymer Light-Emitting Diodes or OLED.
- On OWC research activities there is a European research project action IC1101 OPTICWISE of the COST Programme (European Cooperation in Science and Technology) funded by the European Science Foundation, allowing the coordination of nationally funded research on a European level. The Action aims to serve as a high-profile consolidated European scientific platform for interdisciplinary optical wireless communication (OWC) research activities. It was launched in November 2011 and will run until November 2015. More than 20 countries represented.
- The consumer and industry OWC technologies adoption is represented by the Li-Fi Consortium, founded in 2011 is a Non-profit organization, devoted to introduce optical wireless technology. Promotes the adoption of Light Fidelity (Li-Fi) products.

- An example of Asian awareness about OWC is the VLCC visible light communication consortium in Japan, established at 2007 in order to realize safe, ubiquitous telecommunication system using visible light through the activities of market research, promotion, and standardization.
- In the USA there are several OWC initiatives, including the "Smart Lighting Engineering Research Center", founded in 2008 by the National Science Foundation (NSF) is a partnership of Rensselaer Polytechnic Institute (lead institution), Boston University and the University of New Mexico. Outreach partners are Howard University, Morgan State University, and Rose-Hulman Institute of Technology.

Li-Fi

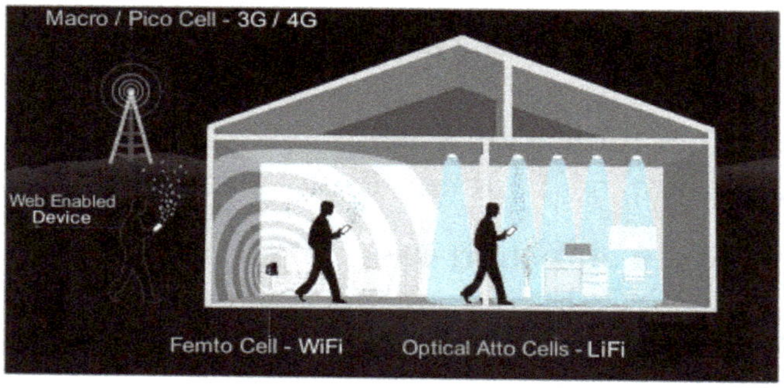

LiFi works in complement with existing and emerging wireless systems.

Light Fidelity (Li-Fi) is a bidirectional, high-speed and fully networked wireless communication technology similar to Wi-Fi. The term was coined by Harald Haas and is a form of visible light communication and a subset of optical wireless communications (OWC) and could be a complement to RF communication (Wi-Fi or cellular networks), or even a replacement in contexts of data broadcasting.

It is wire and uv visible-light communication or infrared and near-ultraviolet instead of radio-frequency spectrum, part of optical wireless communications technology, which carries much more information, and has been proposed as a solution to the RF-bandwidth limitations.

Technology Details

This OWC technology uses light from light-emitting diodes (LEDs) as a medium to deliver networked, mobile, high-speed communication in a similar manner to Wi-Fi. The Li-Fi market is projected to have a compound annual growth rate of 82% from 2013 to 2018 and to be worth over $6 billion per year by 2018.

Visible light communications (VLC) works by switching the current to the LEDs off and on at a very high rate, too quick to be noticed by the human eye. Although Li-Fi LEDs would have to be kept on to transmit data, they could be dimmed to below human visibility while still emitting enough light to carry data. The light waves cannot penetrate walls which makes a much shorter range, though more secure from hacking, relative to Wi-Fi. Direct line of sight is not necessary for Li-Fi to transmit a signal; light reflected off the walls can achieve 70 Mbit/s.

Li-Fi has the advantage of being useful in electromagnetic sensitive areas such as in aircraft cabins, hospitals and nuclear power plants without causing electromagnetic interference. Both Wi-Fi and Li-Fi transmit data over the electromagnetic spectrum, but whereas Wi-Fi utilizes radio waves, Li-Fi uses visible light. While the US Federal Communications Commission has warned of a potential spectrum crisis because Wi-Fi is close to full capacity, Li-Fi has almost no limitations on capacity. The visible light spectrum is 10,000 times larger than the entire radio frequency spectrum. Researchers have reached data rates of over 10 Gbit/s, which is much faster than typical fast broadband in 2013. Li-Fi is expected to be ten times cheaper than Wi-Fi. Short range, low reliability and high installation costs are the potential downsides.

PureLiFi demonstrated the first commercially available Li-Fi system, the Li-1st, at the 2014 Mobile World Congress in Barcelona.

Bg-Fi is a Li-Fi system consisting of an application for a mobile device, and a simple consumer product, like an IoT (Internet of Things) device, with color sensor, microcontroller, and embedded software. Light from the mobile device display communicates to the color sensor on the consumer product, which converts the light into digital information. Light emitting diodes enable the consumer product to communicate synchronously with the mobile device.

History

Harald Haas, coined the term "Li-Fi" at his TED Global Talk where he introduced the idea of "Wireless data from every light". He is Chairman of Mobile Communications at the University of Edinburgh and co-founder of pureLiFi.

The general term visible light communication (VLC), whose history dates back to the 1880s, includes any use of the visible light portion of the electromagnetic spectrum to transmit information. The D-Light project at Edinburgh's Institute for Digital Communications was funded from January 2010 to January 2012. Haas promoted this technology in his 2011 TED Global talk and helped start a company to market it. PureLiFi, formerly pureVLC, is an original equipment manufacturer (OEM) firm set up to commercialize Li-Fi products for integration with existing LED-lighting systems.

In October 2011, companies and industry groups formed the Li-Fi Consortium, to promote high-speed optical wireless systems and to overcome the limited amount of radio-based wireless spectrum available by exploiting a completely different part of the electromagnetic spectrum.

A number of companies offer uni-directional VLC products, which is not the same as Li-Fi - a term defined by the IEEE 802.15.7r1 standardization committee.

VLC technology was exhibited in 2012 using Li-Fi. By August 2013, data rates of over 1.6 Gbit/s were demonstrated over a single color LED. In September 2013, a press release said that Li-Fi, or VLC systems in general, do not require line-of-sight conditions. In October 2013, it was reported Chinese manufacturers were working on Li-Fi development kits.

In April 2014, the Russian company Stins Coman announced the development of a Li-Fi wireless local network called BeamCaster. Their current module transfers data at 1.25 gigabytes per second but they foresee boosting speeds up to 5 GB/second in the near future. In 2014 a new record was established by Sisoft (a Mexican company) that was able to transfer data at speeds of up to 10 Gbit/s across a light spectrum emitted by LED lamps.

Standards

Like Wi-Fi, Li-Fi is wireless and uses similar 802.11 protocols; but it uses visible light communication (instead of radio frequency waves), which has much wider bandwidth.

One part of VLC is modeled after communication protocols established by the IEEE 802 workgroup. However, the IEEE 802.15.7 standard is out-of-date, it fails to consider the latest technological developments in the field of optical wireless communications, specifically with the introduction of optical orthogonal frequency-division multiplexing (O-OFDM) modulation methods which have been optimized for data rates, multiple-access and energy efficiency. The introduction of O-OFDM means that a new drive for standardization of optical wireless communications is required.

Nonetheless, the IEEE 802.15.7 standard defines the physical layer (PHY) and media access control (MAC) layer. The standard is able to deliver enough data rates to transmit audio, video and multimedia services. It takes into account optical transmission mobility, its compatibility with artificial lighting present in infrastructures, and the interference which may be generated by ambient lighting. The MAC layer permits using the link with the other layers as with the TCP/IP protocol.

The standard defines three PHY layers with different rates:

- The PHY I was established for outdoor application and works from 11.67 kbit/s to 267.6 kbit/s.
- The PHY II layer permits reaching data rates from 1.25 Mbit/s to 96 Mbit/s.
- The PHY III is used for many emissions sources with a particular modulation method called color shift keying (CSK). PHY III can deliver rates from 12 Mbit/s to 96 Mbit/s.

The modulation formats recognized for PHY I and PHY II are on-off keying (OOK) and variable pulse position modulation (VPPM). The Manchester coding used for the PHY I and PHY II layers includes the clock inside the transmitted data by representing a logic 0 with an OOK symbol "01" and a logic 1 with an OOK symbol "10", all with a DC component. The DC component avoids light extinction in case of an extended run of logic 0's.

The first VLC smartphone prototype was presented at the Consumer Electronics Show in Las Vegas from January 7–10 in 2014. The phone uses SunPartner's Wysips CONNECT, a technique that converts light waves into usable energy, making the phone capable of receiving and decoding signals without drawing on its battery. A clear thin layer of crystal glass can be added to small screens like watches and smartphones that make them solar powered. Smartphones could gain 15% more battery life during a typical day. This first smartphones using this technology should arrive in 2015. This screen can also receive VLC signals as well as the smartphone camera. The cost of these screens per smartphone is between $2 and $3, much cheaper than most new technology.

Philips lighting company has developed a VLC system for shoppers at stores. They have to download an app on their smartphone and then their smartphone works with the LEDs in the store. The LEDs can pinpoint where they are located in the store and give them corresponding coupons and information based on which aisle they are on and what they are looking at.

References

- "Future Trends in Fiber Optics Communication" (PDF). WCE, London UK. July 2, 2014. ISBN 978-988-19252-7-5
- Nishizawa, Jun-ichi & Suto, Ken (2004). "Terahertz wave generation and light amplification using Raman effect". In Bhat, K. N. & DasGupta, Amitava. Physics of semiconductor devices. New Delhi, India: Narosa Publishing House. p. 27. ISBN 81-7319-567-6
- Measures, Raymond M. (2001). Structural Monitoring with Fiber Optic Technology. San Diego, California, USA: Academic Press. pp. Chapter 7. ISBN 0-12-487430-4
- Bănică, Florinel-Gabriel (2012). Chemical Sensors and Biosensors: Fundamentals and Applications. Chichester: John Wiley and Sons. Ch. 18–20. ISBN 978-0-470-71066-1
- Williams, E. A. (ed.) (2007). National Association of Broadcasters Engineering Handbook (Tenth Edition). Taylor & Francis. pp. 1667–1685 (Chapter 6.10 – Fiber–Optic Transmission Systems, Jim Jachetta). ISBN 978-0-240-80751-5
- Semaphore Signalling ISBN 978-0-86341-327-8 Communications: an international history of the formative years R. W. Burns, 2004
- Mary Kay Carson (2007). Alexander Graham Bell: Giving Voice To The World. Sterling Biographies. New York: Sterling Publishing. pp. 76–78. ISBN 978-1-4027-3230-0
- Vega, Anna (14 July 2014). "Li-fi record data transmission of 10Gbps set using LED lights". Engineering and Technology Magazine. Retrieved 19, January 2020

- Mary Kay Carson (2007). Alexander Graham Bell: Giving Voice To The World. Sterling Biographies. New York: Sterling Publishing. pp. 76–78. ISBN 978-1-4027-3230-0
- pureVLC (10 September 2013). "pureVLC Demonstrates Li-Fi Using Reflected Light". Edinburgh. Retrieved 17, June 2020

Permissions

All chapters in this book are published with permission under the Creative Commons Attribution Share Alike License or equivalent. Every chapter published in this book has been scrutinized by our experts. Their significance has been extensively debated. The topics covered herein carry significant information for a comprehensive understanding. They may even be implemented as practical applications or may be referred to as a beginning point for further studies.

We would like to thank the editorial team for lending their expertise to make the book truly unique. They have played a crucial role in the development of this book. Without their invaluable contributions this book wouldn't have been possible. They have made vital efforts to compile up to date information on the varied aspects of this subject to make this book a valuable addition to the collection of many professionals and students.

This book was conceptualized with the vision of imparting up-to-date and integrated information in this field. To ensure the same, a matchless editorial board was set up. Every individual on the board went through rigorous rounds of assessment to prove their worth. After which they invested a large part of their time researching and compiling the most relevant data for our readers.

The editorial board has been involved in producing this book since its inception. They have spent rigorous hours researching and exploring the diverse topics which have resulted in the successful publishing of this book. They have passed on their knowledge of decades through this book. To expedite this challenging task, the publisher supported the team at every step. A small team of assistant editors was also appointed to further simplify the editing procedure and attain best results for the readers.

Apart from the editorial board, the designing team has also invested a significant amount of their time in understanding the subject and creating the most relevant covers. They scrutinized every image to scout for the most suitable representation of the subject and create an appropriate cover for the book.

The publishing team has been an ardent support to the editorial, designing and production team. Their endless efforts to recruit the best for this project, has resulted in the accomplishment of this book. They are a veteran in the field of academics and their pool of knowledge is as vast as their experience in printing. Their expertise and guidance has proved useful at every step. Their uncompromising quality standards have made this book an exceptional effort. Their encouragement from time to time has been an inspiration for everyone.

The publisher and the editorial board hope that this book will prove to be a valuable piece of knowledge for students, practitioners and scholars across the globe.

Index

www.ingramcontent.com/pod-product-compliance
Lightning Source LLC
LaVergne TN
LVHW080720210926
848485LV00003B/493

* 9 7 8 1 6 3 9 8 9 3 9 7 3 *